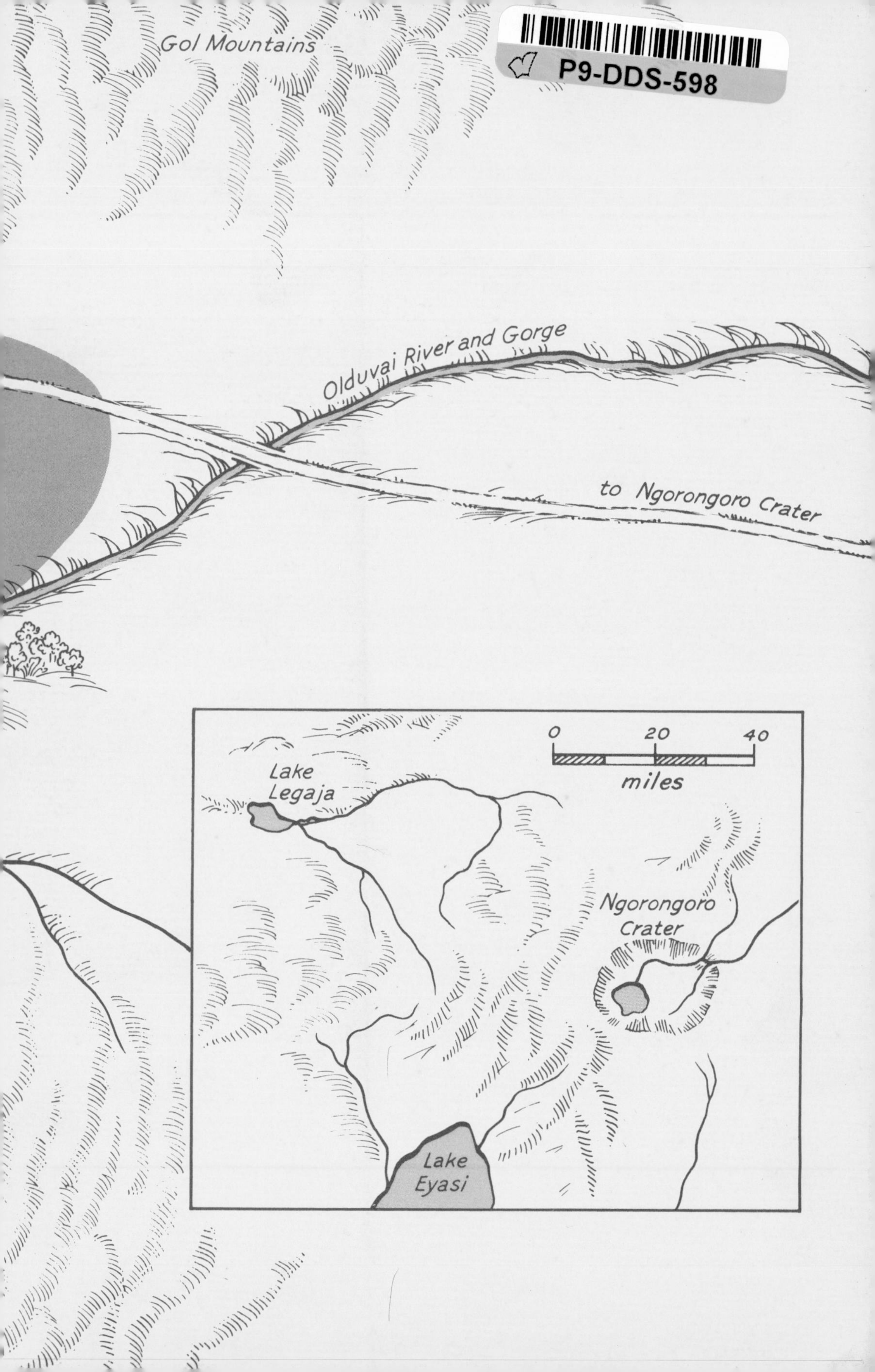
Gol Mountains
Olduvai River and Gorge
to Ngorongoro Crater
0
20
40
miles
Lake
Legaja
Ngorongoro
Crater
Lake
Eyasi

Innocent Killers

Hugo and Jane van Lawick-Goodall

Innocent Killers

Illustrated with photographs

HOUGHTON MIFFLIN COMPANY BOSTON · 1971

First Printing W

First American Edition 1971

International Standard Book Number: 0-395-12109-4
Library of Congress Catalog Card Number: 78-132786
Printed in the United States of America

To Our Mothers

Acknowledgments

We should not have been able to prepare this book without the assistance and co-operation of a great many people and we should like to express here, though all too inadequately, our most sincere thanks to them. Dr L. S. B. Leakey, Curator of Nairobi's Centre for Prehistory and Palaeontology, not only encouraged and stimulated our researches into animal behaviour, but also introduced us to each other. For Louis's timely interference in our separate lives we shall, therefore, always be immensely *grateful.* We should like to thank also Billy Collins, our publisher, who from the beginning has actively shown keen interest and support for our venture. When the wild dogs had young six months later than we had anticipated, Billy was amazingly patient and long-suffering about the delay which this imposed on the completion of the manuscript. Nor, when he visited us a month after the deadline we had failed to meet, did Billy reveal the impatience and concern which he must have been feeling. We are truly grateful to him for his understanding.

We are much indebted to John Owen, Director of the Tanzania National Parks for permission to carry out our research in the Serengeti National Park, and also to the Park Wardens, Sandy Fields and Myles Turner, and the Directors of the Serengeti Research Institute, Dr Hugh Lamprey and Dr Hans Kruuk, for their co-operation and help. We are also very grateful to Ole Saibul, Conservator of the Ngorongoro Conservation Area, for permission to work in the Ngorongoro Crater, and to Dr Pierre Des Meulles and the other staff of the Ngorongoro Conservation Unit for the assistance which they gave us at all times.

For many of the observations in this book we have to thank our student assistants. Unfortunately it proved too confusing to include their names as frequently as we should have wished in the text. In particular, great credit goes to Jean-Jaques Mermod, who worked principally on the wild dogs, and to Benjamin Gray, who worked on the golden jackals. Both were with us for the best part of a year, and both achieved a very high standard of observation and photography. We should like to thank them here for the hard work and long hours which they devoted to our research. We are also indebted to Jeff Schoffern and Roger Polk who helped with the observations on the wild dogs, to Patti Moehlman, Parker Gray and Nicholas and Margaret Pickford, who made many observations on the behaviour of the golden jackal, and to Cathleen Clark who, together with Nick and Margaret, helped for a while with the hyena study. But for these students it would have taken us at least another year to collect sufficient information for this book.

Our thanks are also due to Benjamin Gray for pictures No. 5 and No. 25 in the *Golden Jackal* series, and to Jeff Schoffern for photograph No. 33 in the *Hyena* series: also to Hugo's brother, Baron Godert van Lawick, for taking the photograph of the family on the back of the jacket.

Our acknowledgments would not be complete without thanking, also, our mothers who, during their all too brief visits to Africa, helped us enormously, particularly by looking after our son, Grublin. They assured us that they thoroughly enjoyed this chore, but this in no way lessens our gratitude to them.

Finally we should like to try and thank George Dove. We camped for almost a year close to Ndutu Tented Lodge, the tourist camp which he runs on the shore of Lake Legaja. George not only helped us to find packs of wild dogs, but also repaired our cars, delivered us fresh food, allowed us to share his precious water and, in fact, assisted us in every possible way. He is a true friend, of the sort one finds only once or twice in a lifetime.

Hugo and Jane van Lawick-Goodall

Preface

Jane van Lawick-Goodall's name is very widely known because of her outstanding field studies of chimpanzees in the wild, at the Gombe Stream Park, Tanzania. Her husband, Hugo, is equally well known as the supreme wildlife photographer who was responsible for the magnificent photographs of chimpanzees that have illustrated her articles and books and made her TV and other films on the chimpanzees.

In this new book we have both of them appearing in slightly new roles. Hugo learnt from Jane the technique of animal behaviour study while working with her on the chimpanzees, and he has contributed two chapters of the highest interest – one on the Cape Hunting Dogs and the other on the Golden Jackals. Jane has for the moment turned her talent to the Spotted Hyenas – *Crocuta.*

The uninitiated will wonder what that is new and interesting could possibly be found out about these three species, all of which have been studied by others previously, but Jane and Hugo are not ordinary animal behaviour students, they have a patience and persistence in the field that is almost unmatched, and when this is combined with Hugo's own very special qualities as cameraman, the result – as one would expect – makes most other animal books seem rather superficial.

The present book is planned as one of two describing the larger East African Carnivores, aimed at the general reader as well as animal lovers. Scientific reports giving more detail will also be produced on the basis of the authors' recent work, for the benefit of the specialists in this field.

With such a mass of wholly new information about the three species described, and with such wonderful illustrations, few will be able to resist buying the book, while ethologists and animal behaviourists will require it as a sort of hors d'œuvre for the scientific reports that will follow as the study proceeds.

In the present volume the hyenas are represented by the spotted species, *Crocuta*; jackals by the so-called Golden or Asiatic jackal, while there is, of course, only one species of wild dog – the Cape hunting dogs.

While the publishers are thinking in terms of a second volume to deal with lions, leopards and cheetas, my guess is that the public will demand a third, along the same lines, but dealing with the other hyena species, the rarer striped hyenas, which are wholly different from *Crocuta* and the other jackal species, each with their own very interesting behaviour patterns.

Probably the rarely seen, but not truly rare, nocturnal aard wolf should also be included. If anyone can achieve a really valuable study of the species it will be Jane and Hugo.

Meanwhile the first volume will, I am sure, give huge delight to general public and scientists alike because of its new knowledge and wonderful pictures.

L. S. B. Leakey

Contents

Hunting Grounds

by Day and by Night

A hundred and fifty feet away from us a bull wildebeest galloped, a dark shape against the pale moonlit grass of the African plain. Behind him raced five hyenas, and every moment the gap between the hunted and the hunters lessened. Suddenly the leading hyena seized the wildebeest's tail in its mouth and a minute later the other four were leaping up to bite at the flanks and legs of the victim. Turning swiftly, the bull faced his tormentors, sweeping the darkness with his curved horns, tossing his head. But now more hyenas were appearing out of the night, and within two minutes the wildebeest was down, all but invisible beneath ten or more growling shapes that fought for his flesh.

Hugo drove the car closer to the kill. As he switched on the headlights some of the hunters looked up at us from the feast, their eyes shining, their necks and faces red with blood. Twenty minutes later only a dark trampled patch on the ground remained to tell the story of that moonlit struggle.

That was the first hyena hunt that Hugo and I watched and we were horrified to see, for ourselves, how they ate their prey alive. Since that night we have seen the same gory drama enacted time and time again, for Cape hunting dogs, commonly known as wild dogs, and jackals also kill by this method of rapid disembowelment. We still hate to watch it and yet, though it seems longer at the time, the victim is usually dead within a couple of minutes and undoubtedly in such a severe state of shock that it cannot feel much pain. Indeed, lions, leopards and cheetas, which have the reputation of being 'clean killers', often take ten minutes or more to suffocate their victims, and who are we to judge which is the more painful way to die? And so we do not join the ranks of those who condemn hyenas and wild dogs as vicious brutes that should be ruthlessly exterminated, for they kill in order to eat and to live in the only way for which evolution has fitted them.

It is, in fact, only man who kills with complete awareness of the suffering he may inflict; only man, therefore, who can be guilty of deliberate torture. The history of mankind, if one pauses to think back over the years, is lurid with the so-called inhuman acts of humans and, indeed, the infliction of torture seems to be part of man's heritage. Torture of men and animals alike. And man, it seems, has always been fascinated, in some way, by suffering and death.

Perhaps it is no coincidence that so many observations of Africa's animal killers, the carnivores, are related to their predatory behaviour, particularly to their killing techniques. Hugo and I, however, are interested in more than this. We chose to study them not because they kill but because they are intelligent animals with a fascinating social life. We had watched them and talked about them for years before we started full-time research into their behaviour.

Hugo actually chose photography as a career because he realised that it would enable him to work with animals. When he took his first photograph he knew nothing at all about cameras. He and two friends were touring a National Park in Holland when they came upon a group of shy moufflon sheep introduced from abroad. All three boys wanted to get a photograph. Hugo's companions knew something about cameras – but it was Hugo who knew most about animals. Pushing a camera into his hands, one of his friends told him that diaphragm and distance were pre-set, and that if he could creep to a certain tree all that he had to do was press the button. Hugo took the camera and got his picture. Since then he has developed his talent by trying to photograph animals in action; animals playing and fighting, chasing and being chased, grooming each other and courting each other, feeding and defending their young.

Like Hugo, I began my own work, the study of animal behaviour, in a most unprofessional way. In 1960 Dr L. S. B. Leakey, the well-known palaeontologist who has unearthed so many clues to man's own origins, offered me the chance to study the behaviour of a group of wild chimpanzees at the Gombe Stream Reserve (now Gombe National Park). I accepted eagerly although my only qualification at that time was a life-long interest in wild animals, particularly those in Africa. From the age of eight I had read books on natural history and made my own notes on the behaviour of birds, animals and insects in the area surrounding my home in the south of England. Eventually, although I had never sat for a B.A. examination, I was able to collect enough new information on chimpanzee behaviour to be admitted to Cambridge University to write a Ph.D. thesis.

It was at the Gombe Stream, amongst the chimpanzees, that Hugo and I first got to know each other. It did not take me long to realise that Hugo was no ordinary wild life photographer, but a man who loved and understood animals, a kindred spirit. Dr Leakey had known us both long before we met each other. It did not surprise either of us when we learned, afterwards, that he had written to my mother, before Hugo joined me at the Gombe, to say that he had not only found a young man capable of photographing Jane's chimpanzees, but one who would also make a very good husband for Jane!

After we were married we continued to work together for a year at the Gombe Stream, Hugo collecting an invaluable scientific film record of chimpanzee behaviour under the auspices of the National Geographic Society, which had financed the entire project almost from the beginning. After this it was no longer practical for the Society to support a professional photographer full-time at the chimpanzee camp and they assigned Hugo to photographing animals in East Africa's various National Parks and Game Reserves.

Hugo and I were determined to stay together; we were equally determined that the research into chimpanzee behaviour should go on. We built up facilities for a team of students to continue working at the Gombe Stream, under our overall supervision, and every year tried to spend a few months ourselves working with the chimpanzees. To-day the Gombe Stream Research Centre is one of the few places in the world devoted to long-term research into the social behaviour of one group of wild animals.

When I started to travel with Hugo on his photographic assignment I became more and more aware of the importance, for myself, of studying animals other than chimpanzees. The more I watched the creatures of the Serengeti and other National Parks, the more my own knowledge of chimpanzee behaviour was put into perspective. At the beginning of this phase in our lives, for instance, we made the exciting discovery that the small white Egyptian vulture is one of the few creatures, other than man, to use tools in the wild. At the Gombe Stream some of my most interesting observations had been concerned with the wide variety of purposes for which wild chimpanzees use objects as tools – grasses and sticks for feeding on termites and ants, crumpled leaves as sponges for sopping up water they cannot reach with their lips, leaves for wiping dirt from their bodies, stones and sticks as missiles to throw at either baboons or humans. And now, no sooner had we left the Gombe than we discovered a new tool-user.

We were driving along in the Land-Rover over ground that was black with the aftermath of a grass fire. In places fallen trees still smouldered, sending small curls of smoke into the heat of midday. We had the countryside to ourselves for most of the tourists were having their lunch. Suddenly Hugo spotted some vultures plummeting from the sky in the far distance and we drove quickly over to see what was going on. We found a deserted ostrich nest; some fifteen eggs were scattered about on the ground surrounded by a throng of squabbling vultures. The flames, presumably, had driven the sitting bird from the nest, but miraculously the eggs were scarcely even singed. One hyena ran off as we arrived, and we supposed that the egg over which the vultures were squabbling had been broken open by him. But as we watched, we saw one of the two Egyptian vultures present pick up a stone in his beak and walk over to a nearby egg. Then he raised his head and, with a forceful downward movement, threw the stone at the thick white shell. We could hear the impact. Then he picked up the stone and threw it again and again until the shell was cracked and the contents spilled over on to the ground.

For the next fifteen minutes we watched the two Egyptian vultures opening egg after egg, only to be chased off by the larger vulture species before they had time for more than a beakful of egg. We couldn't speak to each other for Hugo was taking pictures, and I was frantically recording the behaviour on my tape-recorder. Afterwards, when the two tool-users had managed to get enough to eat and had flown away, we stayed watching the other vultures. But though they again and again clawed and pecked at the remaining few eggs, they were completely unsuccessful in opening any and finally gave up and flew away.

Prior to this observation only four creatures had been reported as frequent users of objects as tools in the wild. Two of these were mammals, the chimpanzee and the Californian sea otter. The latter holds shellfish in his paws and cracks them against a stone 'anvil' which he carries up from the sea-bed and lays on his chest whilst floating on his back in the water. The other two were birds, the Galapagos woodpecker finch, which probes insects from crannies in the bark with a cactus spine held in its beak, and the satin bower bird which uses a wad of bark whilst painting its bower with charcoal during courtship.

Hugo and I did a number of tests on Egyptian vultures in different areas and found that in each place they knew how to throw stones. We tested them with smaller eggs too, and found that, rather than break these with stones, they picked up the eggs themselves and threw them against the ground. This, we feel sure, is the behaviour pattern from which the stone-throwing technique has been derived.

Because of our interest in tool-using behaviour, we wondered whether any other animals, with mouths too small to crush the shell of an ostrich egg, would be able to cope with the situation. So far, of the creatures we have tested, only the banded mongoose has managed to solve the problem. And as yet we have not been able to test these mongooses in the wild – possibly our scent on the egg has scared them away. The two tame individuals, however, both of which were caught as youngsters, first tried to throw the egg between their legs against a rock (their normal method of opening small eggs) and then threw stones between their legs at the ostrich egg. Like the Egyptian vulture, the mongoose made use of a behaviour pattern already available to it.

Our new observation on Egyptian vultures was but one indication to us of how little was known about the behaviour of many of Africa's wild creatures. Since then, of course, research into animal behaviour has flourished in East Africa and more and more studies on different species have been started. Many such studies, however, have been concerned with the ecology of the animal concerned, and deal with very large numbers over wide areas. The Serengeti Research Institute, based at Seronera in the Serengeti National Park, has been set up – thanks to the efforts of John Owen, Director of Tanzania National Parks – for this type of research. The combined results of the scientists working there will, in a few years' time, present an invaluable picture of the total ecology of this most famous of all East African parks.

Hugo and I, however, are, and always have been, interested most of all in the social behaviour of individuals within a species. To get good data and photographs on social interactions and the various aspects of individual animals' lives, it is necessary to spend long periods of time watching one family or group. The first step, of course, is to try and get them used to the presence of human observers – not always as simple as it might seem. For, in order to achieve success, it is sometimes necessary to develop an awareness for the feelings of the animals you are watching. I had already learnt this at the Gombe Stream, for the fact that the chimps had accepted my presence did not mean that they would necessarily tolerate my following them about for long periods of time. After one or more hours, depending on the individual, the chimpanzee might suddenly start to walk faster, looking back over his shoulder at me. This was the signal for me to leave him to his own devices.

One student provided us with an excellent example of what may happen if you push too far an animal's newly gained acceptance of a human. The animals he was studying were very tolerant of him when he started his research, and within a couple of weeks he was able to make observations from a distance of a few yards. Over-confident, he kept with them for hours at a time, following them each time they moved from place to place. Gradually the animals became worried until, two months later, they would not allow him anywhere near them. It took another three or four months of patience before he was able to approach them as closely as he had during the second week of the study.

Hugo and I spent some months watching bat-eared foxes, small dainty creatures, fawn-coloured with large dark ears, dark legs, and dark tips to their muzzles and tails. Sometimes these foxes will at first tolerate a car closely approaching their den, but it is important not to watch them for very long periods to begin with. If you do they will simply curl up and go to sleep. Most people would take this to indicate that the foxes are completely at ease, but, in fact, it usually means that they are uneasy and under stress. The behaviour can be compared to that of the fabled ostrich hiding his head in the sand, and it is, in fact, not uncommon in the animal world. A young captive chimpanzee, presented with a test which it cannot accomplish, when it is frightened and in strange surroundings, may simply curl up and go to sleep on the bare floor. Gavin Maxwell, in his well-known book, *Ring of Bright Water*, comments on the phenomenon. His tame otter, when forced to travel in cars, which he hated, would, after a few moments of frenzy, 'curl himself into a tight ball and banish entirely the distasteful world about him'. The same author observed, on different occasions, an arctic fox, a badger and a common house mouse, all of which had gone into a deep sleep, almost a coma, when trapped. There are actually cases of soldiers, in war time, who have gone to sleep under dangerous conditions – when, for example, surrounded by flying shrapnel and bullets. In a different situation, but also as an 'escape' mechanism, I used to curl up and all but sleep during rain storms at the Gombe when I was caught without a coat, and when the strong wind from the mountain peaks made it seem freezing in that normally tropical environment.

Hugo and I got to know the individual members of one family

of bat-eared foxes pretty well. There were three adults at the den, two of which were females, and five cubs. The different foxes looked very much alike, but at last I found I could distinguish them by the pigmentation pattern of the muzzle, and soon we realised that the five cubs were the progeny of both females. After suckling from one mother they would all rush over and suckle from the other. As the cubs grew older they gradually accompanied the adults farther and farther from the den on insect-hunting forays. One day we saw the entire family, adults and cubs alike, playing with a fully grown male Thompson's gazelle. Like streaks of lightning, the foxes darted towards him and then raced round and round, sometimes circling the gazelle, sometimes running in the other direction. With their tails arched upwards, or undulating behind them, they reminded us of a shoal of swift-moving fishes. The gazelle seemed to enter wholeheartedly into the spirit of the game, for he spun round, tossing his spiked horns, pirouetting as he wheeled to face first one and then another section of the moving ring of foxes that surrounded him. A few times he playfully ran towards one of the foxes which streaked away only to turn and once more race around him. We saw the foxes playing with other gazelles after that, but never with one that joined in with the verve of that first male.

Thompson's gazelles, which occur in thousands on the Serengeti, are one of our favourite animals of the plain. This gazelle, which stands just over two feet high, is a beautiful animal, golden russet above and white below, with a vivid black stripe along each side of its body. As these gazelles stand and graze, their short tails constantly flicker from side to side, and young adults of either sex are always breaking off from the routine of feeding to play with each other, chasing round in wide circles at incredible speed, or sometimes, if they are males, jousting with their graceful slightly curved horns. It always seems to us that these gazelles actually play with cars, for they will race along beside the road at the same speed as you and then, suddenly putting on a tremendous spurt, leap across ahead of you.

When you see an adult male putting all his speed into an

attempt to catch up with a flirtatious female, they appear literally to fly across the ground, and their sudden changes of direction are so fast that often your eyes are left momentarily following the original course. It seems that they must move faster than the fastest predator. Nevertheless, they are a favoured prey of wild dogs and cheetas on the open plains, and when they are grazing amongst trees and bushes they are successfully stalked by lions and leopards. During the gazelle birth season an exceptionally heavy toll is taken of pregnant females, especially females actually giving birth, and their fawns.

For the first days of its life, a Thompson's gazelle fawn will try to escape detection by remaining pressed close to the ground at the approach of a potential predator, whilst the mother runs off to return later. The coat of a youngster is much darker than that of an adult, and the camouflage is superb – many a time we have all but driven over a fawn so well does its colouring blend into that of the plains. Despite this, however, countless numbers of these enchanting youngsters, reminding me of miniature Bambis, are destined to die soon after birth. Leopards, cheetas, wild dogs, hyenas and jackals are their chief dangers, but a lion will make a snack of one if he comes across it, and they are also preyed upon by baboons, the larger birds of prey, servals and caracals, the jackal-sized lynxes of East Africa.

The mother gazelle will chase these smaller predators again and again in defence of her young. We watched one female chase a male baboon, who had seized her fawn, for at least two hundred yards until he was able to leap out of her reach into a tree. Another mother charged a huge Martial eagle every time it swooped down near her motionless fawn until, finally, the bird gave up and flew away. And another time we saw a female drive off a long-legged secretary bird, racing after it as it flew low over the ground and then charging as it landed so that it was forced to take to the air again. Often smaller birds of prey that land too close to a fawn will be chased away, as though the sight of any curved beak and talons too close to her young triggers off the mother's defensive mechanism.

It seems strange that the female Thompson's gazelle is in the evolutionary process of losing her horns. That is why you so often see a female with crossed horns, one horn sticking forwards or backwards, one horn missing, or even no horns at all. Certainly we have never seen a predator injured by a mother gazelle's horns, but jackals and birds of prey show respect for these sharp little spikes, and it is difficult to imagine what adaptive value can lie in their gradual loss.

The wildebeest has always been another of my favourite animals. With his long-shaped narrow face and crown of upcurved horns, his fringe of pale hair from chin to chest and his limp black mane, he presents a clownish appearance that is often matched by his behaviour. I never tire of watching the mock-fighting of the bulls before the start of the short annual rutting season. Often you can trigger off such a fight simply by driving past two bulls grazing peacefully side by side. As you approach they suddenly leap around to face each other, tossing their heads and, perhaps, pawing at the ground with one foreleg. One or both may drop to his knees and rub first one and then the other horn in the dirt, after which they usually stand again, facing each other and looking quite peaceful until, as at some prearranged signal, they lunge towards each other, dropping on their knees to the ground and bringing their horns together with a loud clash. The bout may end quickly, or it may be prolonged, with the stronger moving forward on his knees, the loser backing away until he can take no more, then jumping to his feet and running off. Sometimes the victor trots after him, taking long strides with stiffly extended legs and seeming almost to hover in mid-air between each step.

When a wildebeest is in particularly high spirits he may, for no apparent reason, suddenly start to gallop and buck his way through the herd, tossing his head and kicking his hind legs in the air. Often another bull will give chase, and this may end in a mock battle. Best of all for entertainment value is to watch one of these bulls when he pirouettes around his 'rival', leaping high off the ground as he turns grotesquely in the air like some clown pretending to demonstrate ballet.

In the rutting season the bulls fight more seriously, each trying to twist the other's neck and throw its opponent off balance. But we have never seen either of the combatants with a more serious injury than a slight skin wound on the head or neck at the end of such a battle.

One of the tragedies of the wildebeest world is that during the birth season – usually between December and February on the Serengeti – so many calves in the big herds get lost. It only needs one predator, chasing one calf or adult, to set the whole herd running. Indeed, tourist cars, driving too fast, or low-flying planes, can have the same effect. And then, when things have calmed down again, you are almost certain to see several calves wandering about looking for their mothers. Some are lucky and become re-united, but others are less fortunate, particularly in the huge concentrations of wildebeests that yearly trek across the plains of the Serengeti. As the minutes and then the hours go by the bleating of a lost calf becomes more insistent, and again and again it will approach different cows, nosing into their groins as it tries to suckle. But it is rare indeed for a female to accept a calf other than her own, even if she has lost a calf herself and has udders bursting with milk. The following day the bleats of the orphan are fainter, and on the third or fourth day it will lie down to die – if, indeed, it has managed to escape its enemies for so long.

It is not unusual to see a lost calf wandering about at some distance from the herd, and such a youngster often tries to adopt almost any moving object. Hugo and I saw one such calf that kept approaching four hyenas. Each time it got within thirty feet or so, the hyenas started towards it and then, as some instinct asserted itself, the calf ran off a short distance. The hyenas, obviously, were not hungry, for when the calf ran, they stopped following. And then the calf turned and once again approached them. It was a nerve-racking situation, and presently we drove a little closer. The calf suddenly became aware of our moving car and, bleating loudly, came towards us. If we moved, it followed. This is, in fact, a fairly common situation – I always wish it was possible to set up a huge orphanage for these abandoned youngsters, for it seems utterly heartless to drive away and leave them to their fate.

On that occasion Hugo and I, who did not at that time realise that it was extremely unlikely that a foster-mother would adopt the calf, decided to try and lead it back to its herd. But there were three herds visible, one about five hundred yards to the south, the other two slightly farther away to the north and east. Which one should we choose? Finally we decided on the nearest and set off towards it. The calf followed and soon we were within fifty yards. When the youngster saw the other wildebeests it started towards them, bleating. The herd was facing us, as though watching, and suddenly a cow came running towards us, calling loudly. The calf galloped up to her and they nosed each other for a moment before the cow turned and, closely followed by the calf, moved back and was lost to sight in the milling herd. Had we been incredibly lucky and managed to restore the lost calf to its rightful mother? Or had we witnessed one of the rare adoptions? We shall never know, but the cow's behaviour suggested that she was, indeed, the mother, and we drove away feeling pleased.

It was just after that, however, that an incident occurred with a very different ending. We were driving along near the Seronera River when suddenly, round a sharp bend, we came upon a zebra mare who had just given birth. As she saw us she scrambled to her feet and ran off, leaving the foal struggling on the ground. Quickly Hugo reversed and we drove off a hundred yards or so. We saw the foal manage to stand up, tottering, and free itself of the birth sac. But the mare, who had joined a small group of zebras standing some sixty yards away, made no move to return. We drove off even farther, to a point where, with binoculars, we could just see the foal, but the mother, after ten minutes or so, turned round and wandered right away. We were, and still are, puzzled by her behaviour. Possibly it was her first foal and we had disturbed her before she had licked the birth fluids which, according to some scientists, is an important step in establishing the mother's attachment to her young.

Hoping that the mother might return, we drove away, but

when we returned, four hours later, the foal was still on its own. It had 'adopted' a fallen tree and, every so often, tried to suckle on a small projection under the trunk. We saw many zebras pass by, but the foal left its tree 'mother' for none of them. As darkness fell we left it there, still trying to suckle, and I must confess that neither Hugo nor I slept much that night. In the morning the foal's body was being eaten by two male lions, and we could only hope that its death had been quick and painless.

Under normal circumstances the zebra is born into a tight-knit family group – a stallion with his mares and some of their progeny. Unlike a male gazelle or wildebeest, a zebra stallion will try actively to defend his group from predators such as hyenas and wild dogs. Moreover, if one family group is chased, particularly at night, it usually joins up with other groups, until two hundred zebras or more form a united group, many of the stallions staying in the rear and viciously kicking or biting at the predators. This means that, on many occasions, the selected victim, usually a mare or foal, escapes with its life.

In another respect too, the zebras demonstrate a well-developed social co-ordination, which enables all members of the group to sleep soundly for at least part of the night. Hans Klingel and his wife, who studied zebras for seven years, found that the members of one or more families will lie close together. Whilst most of them sleep, some stand guard and quickly alert their sleeping companions should a predator appear on the scene. We watched one such group on a moonlit night and were impressed by the obvious alertness of the sentinel, and the apparently relaxed slumber of the others. We were again struck by the effectiveness of the zebra's stripes as camouflage in the moonlight. On the open plains, in daytime, the zebra stands out clearly, but at dawn or dusk, or when the moon is shining, he becomes almost invisible – in sharp contrast to the conspicuous black hulk of the wildebeest in dim light.

Possibly it is because zebras are able to sleep so soundly at night that you sometimes find one in the daytime stretched out still sound asleep. So sound asleep, indeed, that, on two occasions, we presumed they were dead, for the rest of the herd had galloped noisily away, leaving them motionless on the ground. Only when we got to within a couple of yards did the zebras suddenly scramble to their feet, look around wildly, and gallop after their companions. In the daytime, when sleeping is an individual rather than a group activity, we have not seen sentinels; and perhaps a special signal, given by the sentinels, is necessary to rouse a zebra easily from a deep sleep.

During those two years Hugo and I gradually became more and more interested in the carnivores. We were interested in hunting techniques because I had found that, in the Gombe Stream area anyway, the chimpanzees were efficient hunters and killers of quite large mammals, such as young bushbucks and monkeys. And we were interested in the scavengers, the hyena and the jackal, because many people believe that prehistoric man was a scavenger before he was a hunter. The Gombe Stream chimpanzees will not touch animal flesh unless it has been killed by a group member and, when these apes set out hunting, they are often extremely successful. Now, with the panorama of predator, prey and scavenger spread before us, Hugo and I tried to imagine prehistoric man, whose behaviour might well have been similar, in some ways, to that of the chimpanzees, surviving as a scavenger. For several reasons we found it difficult to believe.

Let me present a general picture of some of the problems which face the hyena and jackals to-day in their scavenging activities. Food which may be scavenged consists of the carcasses of animals which have died a natural death, the remains of the prey of carnivores or offal and so forth around human settlements. One problem for the scavenger is to find such food, which he may do by sight, hearing or smell; a second, if the real killer is still finishing his meal, is to get a share without being hurt; a third is to get there quickly before too many other scavenging competitors have arrived at the scene. And I should say here that scavenging is by no means confined to hyenas and jackals. Lions, leopards, cheetas and wild dogs, as well as many of the smaller carnivores, will feed readily from carrion or try to appropriate the prey of smaller predators.

The hyena is, in many ways, well adapted for a scavenging role. He has enormously strong teeth and jaws, and when, as is so often the case, little remains of the prey animal, is able to chew and digest extremely large bones and tough hide. In addition, he has most sensitive ears and can accurately locate from far away sounds made by other carnivores as they squabble at a kill. He can run at thirty m.p.h. or faster and he has great stamina. He also has patience; a group of hyenas will hang around a lion kill for eight hours, or maybe longer, when they must know from experience that little will be left of the carcass when the killers finally move away.

The jackal also has good hearing, but his chief asset appears to be his speed, which enables him to dart in and seize pieces of meat from under the very nose of a lion or other large predator with little risk of being caught. However, neither the hyena nor the jackal is purely a scavenger, except in some areas, around the habitation of man, where most wild animals have been exterminated and a hyena may exist almost entirely on offal, and so forth. In the crater and on the Serengeti the hyena is an efficient hunter and killer in his own right, and the jackal spends a far greater proportion of his time hunting insects and rodents than in scavenging.

It is only the winged scavengers, the vultures, the Marabou storks, and some of the eagles, that can be considered really efficient. Not only are they able to cover large distances through the air with relatively little effort, but they can maintain vantage points up in the sky which enable them to encompass, with their keen eyesight, large areas of the surrounding countryside. Once they have spotted a dead animal or a predator on its kill, their wings enable them to reach the spot much faster than any four-footed mammal. Indeed, it is by closely watching the movement of vultures in the sky that many earth-bound predators are directed to new sources of food.

Now let us consider early man in the role of a scavenger. He may have been a reasonably fast runner, although, as he had not long adopted an upright posture, we cannot be sure. Undoubtedly he had good powers of endurance, but even though his ears were certainly much sharper than those of man to-day, at least of 'civilised' man, it is most unlikely that they were as sensitive as those of jackals or hyenas. Early man, of course, would have been able to watch the sky for the tell-tale movement of vultures, and could have run to the scene of the kill along with the other scavengers. If he had found only vultures, or perhaps a couple of cheetas or hyenas – or even a single lion – he might have been able to drive them from their meal and appropriate it for himself. But in those early days when man became a flesh eater, his weapons were probably nothing more than rocks, such as the chimpanzee throws to-day. It is unlikely that a small group of men (and it is thought that they did use to hunt in small groups) could have driven a pride of lions or a large group of hyenas from their prey. If, like the hyena to-day, man had had to wait until the hunter itself had finished with its kill, he could, indeed, have cracked the bones and eaten the marrow, but could he have digested hide, or bone itself, in the way the hyena can? It seems unlikely.

Finally, man should be considered in the light of his primate origins. The chimpanzee, as I have said, eats flesh – at some time of the year in fairly large amounts – but we have never seen them scavenging. Baboons also eat meat in many places, and may occasionally try to snatch a share from the chimpanzees. But, if they scavenge, it must be rarely indeed. We have seen countless kills within range of a number of different baboon troops, and never have baboons formed part of the attendant scavenger group. Moreover, primates are, with very few exceptions, strictly diurnal and afraid to move about after dark. Yet it is during the night that a high percentage of kills are made, and it is at this time that the hyena and the jackal get the most out of scavenging – when primitive men would, undoubtedly, have been huddled together asleep.

I am not trying to say that early man never scavenged. Man is, and undoubtedly always has been, an opportunist. Of course, to supplement his newly acquired taste for meat, these stone-age men would have scavenged when the reward was worth it and the risks not too great. We think, however, that it is more likely that man acquired his taste for flesh, like the chimpanzee and the baboon, by hunting small creatures for himself. During the birth season calves and fawns are easy prey if the hunter can manage to outwit the mothers. We have seen adult animals so injured that a group of early men would have had little difficulty in overcoming them.

This, then, initially pinpointed our attention on the behaviour of the larger carnivores. Soon, however, we became fascinated more by the animals themselves, by their individual characters, their obvious intelligence. We found that it was often possible to recognise individuals not only from their colouring patterns but also from traits in their behavior. We always named animals that we were watching once we were certain that we knew them. Some scientists maintain that it is more correct to identify animals by numbers – since we are basically concerned with differences between individuals, we found names more satisfactory and equally scientific. I had done the same when I began studying chimpanzees six years earlier.

When Hugo began to plan his long-term study on these carnivores, I never realised that I should become seriously involved in the venture. My own work of analysing and writing up my data on chimpanzee behaviour, together with looking after our newly born son, Hugo junior, more commonly known as Grublin, would, I thought, take up all my time. Yet I think that Hugo, right from the start, knew me well enough to be sure that, in the end, I would join him. Our partnership is such that we work at everything in life together, from writing books to putting nappies on our baby.

It did not take much to persuade me that hyenas are second only to chimpanzees in fascination – they are born clowns, highly individualistic and live in an extremely complex and well-ordered society. But it was difficult for me to conceive, in those early days, that I would be able to study them, handicapped as I was with a baby. Nor *was* it easy when Grublin was small, but fortunately hyenas are most active during the hours of darkness, and so, during the brilliant moonlight of the African nights, I was able to spend many hours watching them whilst Grublin slept peacefully behind me on the big bed in our Volkswagen bus.

I still remember clearly the day we arrived at Ngorongoro Crater, in Tanzania, to begin this work. We had left our house just outside Nairobi the day before and had spent the night camping on the Serengeti plains. In the morning we started out on the last lap of the three-hundred-mile journey. As we drove, in first gear, up the steep slopes of Ngorongoro it grew colder and colder, and eventually we were driving through the thick clouds that hung low over the mountain. When we reached the rim of the crater, or *caldera* as it should be called, we stopped to give our nine-month-old son a drink. As soon as Hugo switched off the engine we became part of the ghostly world through which we had been driving. The white gently moving mist of the clouds closed around the car and all that we could see were a few shrouded outlines of trees and the tall wet grasses at the side of the track. On either side of us, we knew, the thickly forested slopes dropped down; on the one hand to the rolling miles of the Serengeti plains, on the other to the deep basin of the crater itself. The magnificence of the wild country that stretched out below was completely hidden from us by the mist; had we been merely passing tourists we should have missed for ever that fantastic view. Just as, had our lives run on different courses, we should never have learned of the vivid personalities of Mrs Brown, the old hyena mother, or Jason the golden jackal, or Ghengis the leader of the wild dog pack. Yet they, like the view, were there down below the cloud masses, living their lives, sleeping and playing, hunting and killing, mating and giving birth to others of their kind.

Later, as we drove down to the floor of the crater, we left the dense clouds behind, clustering around the rim, and through the thinning mist the green plain below us began to appear. From that height the one hundred square miles of the crater floor seemed, as it always does, quite empty of animal life. Nothing could be farther from the truth. The dark masses of the wildebeest herds and the single black spots of solitary rhinos were the first to stand out clearly from the background of the plains; then we made out groups of zebras and finally the pale sandy-coloured herds of Grant's and Thompson's gazelles. Down below us the small soda lake was fringed with the pale pink of the flamingoes that fed in the shallows, and in the little Lerai forest behind the lake there were, we knew, elephants and buffaloes, baboons and monkeys. Grazing the taller grass of the rolling hilly country on the far side of the crater there would be more elephants and buffaloes and huge herds of elands, Africa's largest antelopes. Hugo and I were setting out to study carnivores, and down below us were carnivores of many species. In the actual basin of the crater several prides of lions were resident, some of the males sporting the magnificent black manes for which the lions of Serengeti are so famous. The spotted hyena is as plentiful in the crater as anywhere in Africa – indeed, until a few years ago the Game Wardens used to shoot fifty or more of them a year to keep the numbers down. To-day they are no longer molested by man and, possibly because of the resulting increase in their population, the smaller wild dogs and the cheetas which used to be found on the crater plains have, for the most part, moved their hunting grounds elsewhere.

All three species of Africa's jackals can be found in the crater basin, the Asiatic or golden jackal, the silverback or blackback jackal, and the rarely observed side-striped jackal. The bat-eared fox, smaller than the European red fox and, in fact, not a true fox at all, is plentiful on the plains, and the exquisitely beautiful serval, a jackal-sized member of the cat family, frequents the tall grassland along the rivers and in the hills. The leopard inhabits the forested slopes, and occasionally can be seen in the crater basin itself, and there are a number of the smaller carnivores in the forests and on the plains; the East African wild cat, the civet, the dainty genet with its long banded tail, and several types of mongoose.

We drove slowly to the log cabin at our camping place on the far side of the crater plain. Everywhere the grass was lush, and we passed literally thousands of wildebeests. It may seem strange, to some, that I write wildebeest*s*, using the plural. Most people will talk about a herd of wildebeest, or zebra, a pride of lion, and so forth. But to us, this use of the singular suggests that the individuality of each animal in the group is being ignored. It implies, to us, that every lion is just a lion. After all, who would dream of talking about a boatload of Italian, a classroom of German, or even a gathering of man? And so, quite deliberately, Hugo and I refer to a group of animals in the plural.

As we passed the herds our son got more and more excited, almost jumping out of the window in his efforts to get closer to the animals. Occasionally a jackal pricked up its ears to watch our approaching car before jumping up and running a short way from the track. Once a fat ungainly hyena heaved herself from the muddy ditch by the side of the track and loped off, looking back, hyena fashion, over her shoulder.

The cabin was 'home' as soon as we set foot inside it. Built under the shadow of a giant fig tree, it is always cool, surrounded by dim green light, the twittering and chirping of birds, and the constant babbling of a small muddy stream. This stream, known somewhat grandly as the Munge River, has its source up beyond the rim of the crater, and after winding its way down through the hilly country behind the cabin, goes chuckling through the roots of the fig trees that mark its course until it empties into the crater's lake. From the cabin the view, framed by the low hanging branches of the fig tree, stretches across several miles of grass plains to the lake and the Lerai forest, behind which the crater wall rises, cutting off the outside world. The cabin itself is a simple one-roomed wooden building, with a few basic items of furniture – tables, shelves, a cupboard and a large bed. The floor is stone, covered by rush matting, the windows are small, the ceiling low. It was built for a student studying the crater's wildebeest herds, and it is now loaned out, by the Ngorongoro Conservation Unit, to other scientists who want to work in the crater.

A few yards away from the cabin is a tiny bamboo-walled hut which serves as a kitchen, although most of the actual cooking was done by our two African servants on an open wood fire outside.

During our stay at the crater the cabin served principally as a safe nursery for Grublin. Hugo and I slept there with him, and we

1. The drive up the Ngorongoro Crater wall to collect supplies.

2. Jane pauses in her observations on hyenas to feed Grublin.

3. Most of the rhinos in Ngorongoro are very placid.

4. A zebra mare repulses a stallion.

5. Two male Thompson's gazelles in a territorial fight.

6. In February the wildebeest calves were born.

7. Bat eared foxes spend much time grooming each other.

8. Grublin learnt to imitate the sounds of animals, such as wildebeests, before he could talk.

9, 10, 11. Egyptian Vultures break open ostrich eggs by throwing stones at them.

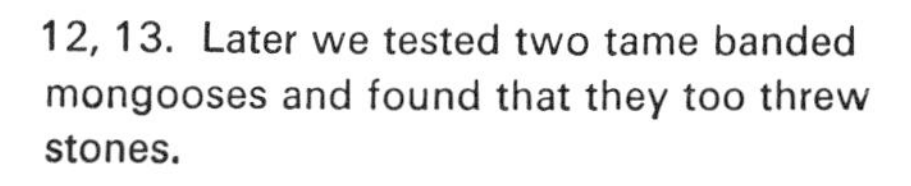

12, 13. Later we tested two tame banded mongooses and found that they too threw stones.

14. A rupell griffon vulture coming in to land.

15. The Sacred Ibis is a common bird in the Crater, where water is plentiful.

16. The marabou stork, like the vulture, is a scavenger.

17. A zebra foal stares from between its mother's hindlegs.

18. We had to collect drinking water from fifty miles away.

MG
16

19. The Ratel, or Honey Badger, is one of the most aggressive animals in Africa.

20. A Serval pounces after a rat over the high grass.

21. Cheetah cubs at play.

erected a number of tents in which to work, eat and so on. Grublin played in the cabin during the daytime, with one of us in constant attendance or, if we were all sitting in the 'dining-room' tent overlooking the plains, he came and crawled around amongst us.

'Surely you are not going to take a baby into the wilds with you?' many of our friends had asked, after Grublin's birth. 'You'll have to change your way of life a bit now, won't you?' said others, with a laugh. But Hugo and I had decided, before the baby was born, that if we could help it we would not allow his arrival to change our life together. Hugo is a wild life photographer and naturalist and so must spend long periods of time in the bush; both of us feel that, whenever it is possible, a husband and wife should be together. When we arrived at the crater Grublin had already spent five of his nine months of life with us in the bush, and it would have been hard to find a healthier, happier baby.

We took precautions, of course. We had with us a radio telephone and could, at any time, ring through to Nairobi. This meant that if Grublin, or any of us for that matter, fell ill or had a bad accident, we could either contact the flying doctor service or charter a light aircraft to fly us to hospital. Whilst we were camped at the crater we never left Grublin alone except when he was sound asleep in the safety of the cabin, and we had a baby alarm set up in the dining-room tent and the working or office tent so that one of us would be sure to hear his first waking sounds. As an added safety measure we put a wire-netting enclosure round the cabin so that, in the unlikely event of his eluding us for a few minutes, he could not stray away.

For about two months during that visit the fig trees were laden with red fruit and, once a day, a troop of baboons came to feast above the cabin. We had to be very careful to keep Grublin inside then, for the big males were insolent fellows and often sat around on the ground eating windfalls and paying scant attention to our presence. On one occasion our tall African cook, Moro, of the Luo tribe, had a narrow escape when a male baboon, which had been fighting with another in the lower branches of the tree, fell heavily, missing him by a matter of inches. Had he been hit the force of the impact would have killed him and, even if it hadn't, the baboon, terrified at the close contact with a man, might well have attacked. The baboon has huge canine teeth and can inflict just as much damage as a leopard. There were, of course, other hazards, not dangerous but unpleasant and smelly, to beset the unwary person who walked beneath the tree when the baboons were feeding overhead.

By far the most dangerous aspect of the Munge camp was the fact that wild animals could approach closely without our knowledge because of the dense vegetation that grows along the river banks. One morning, for instance, Hugo noticed a male lion strolling into the fifteen-foot high undergrowth near the dining-room tent. We drove in with the Land-Rover to try and move him on – and discovered six lionesses lying up there, as well as the male. One of them was only about ten yards from the forty-four-gallon drum of petrol from which Hugo and Moro had just been filling the Land-Rover. We tried to drive them out, but they only moved closer and closer to the stream where we could not follow with the car. For the rest of that day, whilst Hugo was out on the plains, Grublin and I had to stay in the cabin, whilst Moro and his assistant, Thomas, chose to remain up in the fig tree, keeping a lookout until Hugo returned in the evening. By the next morning the lions had gone and they did not return.

There was another incident when Hugo had a narrow escape. About fifteen yards behind the cabin is a tiny bush lavatory, or 'choo'. This is simply a hole in the ground with a wooden box on top of it as a seat, surrounded by a ramshackle circular grass wall with a gap on the far side serving as the entrance. On one side of the narrow path leading to the choo is high undergrowth; on the other side is a steep drop to the stream. Hugo walked blithely along the path and was just about to turn into the entrance when, almost subconsciously, he became aware of something yellow through the dilapidated wall. He paused for a moment, and that may have saved his life. There was a deafening roar and a rending and crashing sound as some large animal broke through the wall on the other side and forced its way through the undergrowth.

Rushing back along the path, Hugo reached the safety of the cabin and, looking out, saw a lioness standing staring back, her mouth slightly open in a snarl and her tail lashing angrily from side to side. She must have been lying up inside the choo; later we saw the great holes in the floor where her paws had broken the ancient wooden boards as she took off to leap through the wall. Why was she there? We soon realised when Hugo spotted the remains of her kill draped over the base of a tree trunk at the edge of the stream, just below the choo. We dragged it away, for we did not relish the idea of the lioness hanging around. That evening Grublin and I and the two Africans watched when the huntress returned to finish off her kill. She stood looking round, her tail lashing, for some ten minutes before stalking slowly past the cabin and moving off to seek her dinner elsewhere.

But life on safari is very rarely spiced with drama of this sort. The hazards involved are no greater than those which people in more civilised countries face every day when they drive on the busy highways. We ourselves feel that, provided we never permit ourselves to be lulled into a sense of false security and provided we are constantly alert to the possibility of attack from a wild animal, Grublin will be as safe as if he were brought up in an English town. We have never had cause to regret our decision to bring our son up in the bush, to share with him some of the experiences we have in our work with wild animals.

During our first months at Ngorongoro, Hugo's mother, known to us all as Moeza, came to stay with us. She was of tremendous help in looking after Grublin. Nor did Hugo lack for assistance, for at that time we had three students with us who had volunteered to help for a while, Parker and Ben Gray and Patti Moehleman. Between them they put in hundreds of hours of observation on the golden jackal family that Hugo was studying.

This study started off gloomily because the grass of the crater plains, normally a couple of inches high at that time of year, was a foot high in many places. The jackal is no bigger than a European red fox and this high grass would make photography and even observation difficult. Miraculously, though, Hugo was lucky, and found a den in an area where the grass was much shorter than in most places. True, there had been a few long blades, most irritating to a photographer, but these were eliminated during the heat of one midday when the four tumbling cubs were down their burrow, and Jason and Jewel, their parents, were out hunting.

Our original plan had been to stay in the crater for about three months, move to the Serengeti for a while to start our wild dog study and then return to Ngorongoro in September to continue work on the jackals and do some intensive observation on the hyenas. But Africa is still close enough to nature for even the best laid plans to be overthrown. On this occasion it was the rain that lengthened our first visit to the crater into nearly six months. The short rains, from November to January, had been unusually heavy; the long rains, lasting until April or May, were even heavier. By the end of March the Munge River had flooded several times and the level of the lake had risen dramatically. Over much of the crater plains the sun glinted on to flood waters; it was quite impossible to move our camp.

Hugo, Parker and Ben (for our third student helper had left us by this time) were often hard put to it even to get to the jackal den, and spent many weary hours digging the car out of one muddy pot hole after another. Indeed, for nearly a month we were completely cut off from the outside world, for both routes to the crater rim were flooded. Had there been an emergency we could have got out – but only by walking and leaving all our equipment behind.

After the rains have stopped it is not long before the tropical sun begins to dry the grass, particularly the short lush grass of the crater plains. Then many of the herbivores move to graze the hilly country to the east of the crater basin. Usually the zebra herds are the first to move because they are adapted for grazing long grass: they are followed by the wildebeest, and when they, in turn, have grazed the grass shorter still, the herds of Thompson's and Grant's gazelles move there to feed.

The animals of the crater are by no means confined within the bowl. Often whole herds migrate, moving up the steep slopes of the wall in long lines, following well-worn animal trails. The country beyond the crater comprises thick forest, mountain ranges and open grassy plains, and is sparsely populated with nomadic groups of the Masai tribe. They are a splendid people, upright of stature, with finely chiselled features, pale coppery-coloured skin, and a tribal heritage that defies the softening influences of Western civilisation. They roam the plains and mountain slopes, as their forbears have for generations, grazing their herds of cattle, sheep and goats alongside the wild animals. The Masai have a tradition of fearlessness which is well deserved. In the old days, before it was forbidden by law, a Masai youth could not expect to marry until he had taken part in a lion hunt, armed only with his spear and shield. Whilst the conservationist must condemn this ancient custom, anyone who has seen a lion charge will recognise the almost incredible bravery which must have been shown by the young men who attempted the feat.

Hugo and I have become friendly with a number of Masai warriors, and we have great admiration for the tribe; as well as being fearless, they are friendly, generous and gentle and loving with their children. Many of them, too, have a fund of knowledge about the countryside and the wild animals, as one would expect of a people living so close to nature, and we always make a point of driving to one of the villages, or manyattas, when we are searching for rare animals in the vicinity. Often they have given us invaluable information.

To the north-west of Ngorongoro the short grass plains stretch for hundreds of square miles, gashed by the twenty-mile long Olduvai Gorge, made famous by the excavations of Dr Leakey and his wife. There *Zinjanthropus* (nicknamed Nutcracker man) was found and, later, *Homo Habilis*, fossilised along with his stone tools and the remains of the animals which he hunted. Here, too, Dr Leakey uncovered the foundations of what was, undoubtedly, one of the earliest walled huts built by man. Beyond Olduvai the plains stretch on, scarcely a tree to be seen as the miles go by. Near the entrance to the famous Serengeti National Park the short grass gradually gives place to longer, as the soil changes in composition, but the plains still stretch ahead until, some sixty miles from Ngorongoro, the Seronera River winds its way through the acacia trees of the Seronera Valley.

The southern boundary of the Serengeti National Park makes a hairpin loop to include, within the park, the small soda lake, Lake Legaja. It was here that we set up our second 'home' in the bush, a camp under shady acacia trees overlooking the water. The lake is also known as Ndutu. Both names, in Masai, have a very similar and most delightful meaning, implying that the place is a peaceful one, sacred to God, and should not be desecrated by the noise of people. How one word can imply so much I cannot imagine, but that is what we were told.

The lake lies within a thin band of acacia trees and thorn scrub beyond which the windswept treeless plains stretch for miles. When we arrived at this new camp site to start concentrated work on the wild dogs, it was February and the plains were thick with the herds of the wildebeest and zebra migration. Nowhere in the world to-day can one see wild animals in such teeming multitudes as on the Serengeti plains during the annual trek of the herds across the newly-green countryside. The movement starts with the coming of the rains: the herds, which have been scattered throughout the bush country to the north and west of the park, close to permanent water, then congregate and, in a number of long dense columns, gradually graze their way to the short grass plains where they remain to calve and feed until the end of the rainy season. Then, as the surface water dries up, the herds once again move back to the bush country and disperse. A total of more than a million herbivores take part each year in this migration, of which half are Thompson's and Grant's gazelles, some 350,000 are wildebeests and some 180,000 are zebras.

For a few weeks, while the herds remained in the vicinity of our camp, we lived our lives to the constant accompaniment of the mellow lowing and honking of the wildebeests and the wild bursts of zebra calls which sound somewhat like speeded-up and hysterical versions of donkeys braying. The splendour and the freedom of those hundreds of miles of unspoilt country, the sunrises and sunsets over plains made black by thousands of animals, the roaring of lions and the weird whooping calls of hyenas at night, are things which I shall remember as long as I live.

For those carnivores living on the plains the arrival of the migration each year heralds a period of plenty. And, whilst there are some lions, cheetas, hyenas and jackals that seldom, if ever, leave their own well-marked territories, there are others of the same species which take advantage of the good living provided by the migration, and which follow the herds over at least part of their yearly route. Often, in the thick of the migration, the eater of flesh does not even need to hunt for himself. There are so many wildebeests and zebras that natural deaths are commonplace, and the vultures, as they plummet down from the sky, quickly give away the whereabouts of such an easy-to-get meal. If early man followed such migrations, then, indeed, he would have been able to scavenge for his living for a while.

Even in the midst of plenty, however, a carnivore may starve to death. I shall never forget the crippled lioness we came across, stretched out under a tree only a few yards from our camp. She was so thin that it was hard to believe she could be alive, but when we drove close she wearily raised her head and looked at us out of huge sunken eyes. When the sun sank lower and shone through the leaves directly on to her, she even got up to move into the shade, half hopping, half dragging her crippled back legs behind her. It was obvious that she could never recover and the kindest thing would have been to end her suffering. But we were in the National Park, where there are strict rules that one must not interfere with the course of nature. So we drove away and left her alone.

That night we put the cars even closer to our tents, for the thought of a starving and wounded lioness so close to camp was terrifying. In the morning we could not find her, though we drove back and forth through the acacias and thorn bushes for almost an hour. All that day and the next Grublin was within a stone's throw of one of the cars, and our African staff kept a constant lookout.

The following day Hugo found the vultures tearing at the dead body of the lioness. She had circled right round our camp and, somehow, dragged herself another three hundred yards into the bush.

Normally the migration stays in the Legaja area until late May or early June; the year we were there the long rains failed, there was not enough surface water to support the herds, and they moved off precipitately at the beginning of March. Just as the flooding had thwarted our plans the year before, the threatened drought upset them the next. For three years running Hugo had found wild dogs with pups near Lake Legaja in March and April. But when we set up camp there in February, for the express purpose of studying wild dogs, there was no den. This made Hugo's study extremely difficult, for these dogs, unless they are forced to remain in one place to look after small puppies, roam freely throughout a vast range, seldom staying in one area more than a few days at a time. Whilst the migration of Thompson's and Grant's gazelles, following the wildebeest and zebra herds, grazed the plains around Legaja, Hugo and his two new student helpers, Jean-Jacques Mermod and Roger Polk, were able to observe a number of different packs of these nomadic hunters. Once a pack had been located, the three of them took it in turns to stay with the dogs, until they lost them when the pack moved on moonless nights.

In April, the gazelles followed the rest of the migration, and after that it was seldom that a pack of dogs was sighted, though Hugo, Jack and Roger fanned out, in their three cars, covering between them an area of about five hundred square miles in a day. They were helped, too, by our friend George Dove. He has a tented safari camp at Lake Legaja, and he told all his drivers that we were searching for dogs: if a pack was sighted George sent someone over at once to let Hugo know.

However, despite all the frustrations and worries, Hugo managed to get some fascinating and completely new information on wild dogs and, on the whole, it was a happy camp. Particularly for Grublin. He was an extremely active two-year-old by this time and, as an extra safety measure, we had taken a third African helper on safari with us. Moro, Thomas and Alec took turns in keeping a constant eye on Grublin. My mother was with us too, on a long visit, and George Dove developed a special fondness for our son so that Grublin never lacked for friends. His chief delight, at this time, lay in playing his version of football, and it was a great sight to watch him enjoying a game with Moro and Alec, both of whom are over six and a half feet tall.

Grublin was always excited when animals wandered close to camp as they so often did, particularly when the migration was around the lake. Once, indeed, we all had to jump into the car when two lions bounded between our tents as they chased a baby wildebeest. And even when the migration had moved on we could still watch our resident group of eight giraffes each morning, and the small herd of gazelles that never went too far from the vicinity of our tents. Sometimes, too, an old bull rhino wandered by. In the evenings, when Grublin was eating his supper outside the tent, long strings of graceful flamingoes flew past, silhouetted against the red or golden sky, and giving their strange creaking calls as they headed for a night's feeding on the lake.

Often we took Grublin out in the car to watch animals, but although he loved it I always dreaded that we would come across some shy creature which Hugo wanted to photograph, such as a caracal, a honey badger or the seldom seen striped hyena, all fairly common around Lake Legaja. For then I had to use all my ingenuity to keep Grublin quiet lest he give a loud yell at the critical moment and I was hardly able to look at the animal at all myself. I was very grateful when my mother volunteered to take over responsibility for our son, and so gave Hugo and me rare opportunities to go out together.

One evening, well before sunset, a striped hyena visited our camp. These animals are not uncommon around Lake Legaja, but they are rarely seen, and virtually nothing is known of their behaviour. On this particular evening the hyena, a remarkably handsome individual with dark wavy stripes on a creamy background, wandered past the kitchen tent and paused to look in. I was giving Grublin his bath, but at Moro's soft call I wrapped my son in a towel and hurried out. Suddenly the hyena pricked his ears and began to run. He vanished over the edge of the slope leading to the lake and we followed in the car just in time to see him chasing a serval which had just caught a hare. The small graceful cat ran fast and after a few yards the hyena gave up, stood for a moment, and then with a glance at his human spectators, wandered on his way. On another occasion Hugo and I followed a striped hyena as it foraged at night. It paid little attention to our car as it walked along sniffing at the ground and occasionally pausing to scent-mark a tuft of grass. We watched as it chased after a steinbuck – a small antelope only a little larger than a dik-dik – but its prey escaped. Soon the hyena moved into thick bush and we could no longer follow.

Sometimes we all went for a drive in the dark. Grublin came with us and loved watching for the eyes of the nocturnal creatures gleaming in the headlights. He was most excited by the leaping eyes of the arboreal bushbabies, shining like red Christmas tree lights as the agile little primates leapt through the branches of the trees. He enjoyed, too, the spring hares whose brilliant eyes move in curved arcs through the darkness as their owners jump along like diminutive kangaroos.

In the crater we had had trouble with rats nibbling their way into all our possessions. At Legaja it was the African dormouse which pestered us. But although the damage to our clothing and papers were about the same, it was somehow easier to forgive these dainty tree-living rodents with their big eyes and long fluffy tails. One morning I picked up a jam jar to spread Grublin's breakfast toast and there, crouched on a thin layer of jam at the bottom, was one of these villains. Grublin and I tipped the jar on its side and it was comical to see the dormouse run out, his tail no longer fluffy but clogged and sticky. After licking himself clean, I suspect he kept well away from strawberry jam for the rest of his life.

In June Hugo and I had to go to Europe and, when Hugo returned to Africa after ten days leaving me to attend some conferences, he planned to move our camp away from Lake Legaja. But George Dove greeted him with the news that he had found a wild dog pack with a den, and thought they had pups. Sure enough they had, and so after having given up all hope of getting detailed information on this aspect of their behaviour, Hugo had his chance after all.

When I returned with Grublin in August, I was appalled to see how parched and dry the plains had become, though it was only to be expected as, apart from a couple of storms, scarcely a drop of rain had fallen since February. Everywhere the ground was covered by dried-out spikes or curled wisps of yellow-grey grass. Dust was a constant nightmare, filling our noses and mouths and lungs, covering everything with a grey film, insinuating its way into all but the most carefully sealed camera cases, rising knee high when we walked along the well-worn tracks between our living tents. Yet dust can be amazingly beautiful too, as when gazelles race across the plains, black shapes amidst a haze of dust, golden red in the light of a setting sun.

As the days went by, and the plains became more and more desert-like, it was an ever-growing source of wonder to us that so many different animals could survive. Yet on our drives we nearly always saw giraffes and gazelles, warthogs and ostriches, hyenas and jackals, and a variety of smaller creatures. And the dogs were there too, finding sufficient prey to feed themselves and their pups. Eventually, though, even the dogs left to resume their nomadic wanderings, and so, finally, we packed up our camp and after two years of concentrated work in the field, returned, for a longer than usual spell, to our house just outside Nairobi. Grublin for a while lived the life of a normal little boy; played in the garden, went to a nursery school in the mornings; slept in a cot in a house at night. Perhaps he did not even miss the open spaces where he had spent so much of his short life. But Hugo and I, as we sorted out our photographs and our thoughts, kept wishing we were back in the crater or in our tents at Legaja, back with the creatures we had got to know so well.

Each of the animals which we studied revealed its own individual character, quite different from that of its brother or father or neighbour. This will not be surprising to some animal lovers. A dog owner is usually quick to affirm that every dog has a completely different individuality. I know one woman who, throughout her life, has owned a succession of cocker spaniels: these dogs have not only come from the same kennels and been trained by the same person, but have also been brought up in the same house. Each one, she maintains with pride, has been completely different from every other. There are many people who will tell you the same thing, not only with relation to dogs, but also cats, horses – and even pigs, sheep and cows. Yet, strangely, the owner of a pet wild animal often has a different attitude and is somehow convinced that his pet has acquired its character through its close association with man. Whilst he may realise full well that two different fox cubs, brought up by him, will have two quite different personalities, he seems unwilling to admit that the same creatures, in the wild, can show the same sort of vivid individuality. His pets, by becoming part of the family, become different from others of their kind in the wild. This, I suppose, is why many hunters who have pet animals feel no compunction in shooting wild ones of the same sort.

One of our aims in writing this book has been to try to show that an animal has as much character when it is wild and free as when it is tamed and brought up by humans. Of course, it takes far longer to appreciate the personality of a creature in the wild because the observer cannot interact with it, and many people base their assessment of an animal's character on the way in which it responds to them personally. In our work such judgment can only be reached after long periods of watching and recording. When Hugo first saw Black Angel, the wild dog female, he recognized her because she had lost half of her tail. It wasn't until he had been with the pack for weeks that he came to know Black Angel as an individual, as different from the other wild dogs as my friend's current cocker spaniel is from any of her previous ones.

This book, the first of two dealing with Africa's larger carnivores, is about three of the most maligned and little understood species; they are, nevertheless, three of the most fascinating to watch. It does not surprise us that most people are horrified at the thought of animals which eat their prey alive, but we have made no attempt to gloss over this aspect of their behaviour. Instead we have tried to present as comprehensive a picture as possible, hoping that a better understanding of the creatures and a glimpse into some of their less well-known but interesting and often charming characteristics will show them in a better light. One incident suggests we may not be too optimistic in this hope. A friend of ours, who has spent most of his life farming in East Africa, came to visit us when we were on the Serengeti. Hugo was able to show him the pack of wild dogs – of which Black Angel was a member and which had pups at the time. One evening when we had gone to the bar at the Lodge for a drink, I chanced to overhear a remark our guest made to an acquaintance. 'Well, one thing's for sure. I'll never shoot a wild dog again. I know too much about them.' It was one of the most heartwarming things I had heard for years.

It may, however, take a long time to wear down people's prejudices. Even in Europe people can have the strangest misconceptions about animals: that hedgehogs steal cows' milk; that bats will become entangled in a woman's hair; that Alsatians are always untrustworthy with small children. The only time in my childhood that I remember being really rude to an old lady was when I was in a field, during a treasured country holiday, stroking a pig. He was one of those black and pink Saddlebacks, and it had taken me days of proffered apple cores and potato peels before he would allow me to touch him. The old lady called me imperiously to the fence, only to tell me that I should never touch pigs for their hair would give me unspeakably horrible diseases, and I should not breathe their breath for the same reason.

Small wonder, then, that such little-known creatures as the subjects of our book should come in for their share of disrepute. A little while ago, when we were driving from Nairobi to the Serengeti, the usual popular opinion of the hyena was brought home to us. We had a young Englishman in the car who had wanted a lift. Hugo saw the body of a dead animal on the road, and we peered ahead to see what it was.

'Oh, it's only a foul hyena,' said our travelling companion. 'Good riddance to it.'

And then, before Hugo and I could comment, a small, rather worried voice piped up from beside me:

'Poor hyena all broken. He like Mummy's hyenas. What's happened him?'

Grublin had been brought up with wild animals from an early age. Photographs of hyenas, in our files, are not just hyenas to him. He cannot identify the different individuals, for that takes weeks of training and practice, but he knows they have different names and he will ask what each one is called. He, surely, will develop an appreciation of animals in the truest sense. And, because more and more is being found out about the ways of wild animals, and because animal books are becoming more accurate and factual and less and less filled with fantasy and exaggeration, there is hope that the younger generation of to-day will grow up without many of the fallacies and misconceptions concerning animals which have prevailed for countless generations.

Our research into the behaviour of the carnivores will have been worthwhile indeed if by sharing our knowledge we can kindle in others something of our respect and affection for these innocent killers.

Wild Dogs

Nomads of the Plains

Wild Dogs

Nomads of the Plains

Some five miles from Lake Legaja, eleven wild dogs trotted along in single file. As usual Genghis, the old male, was in the lead. All around, the short grass plains were dry and barren and the small groups of gazelles that grazed there were few and far between. It was August, the middle of the long dry season. Suddenly Genghis slightly altered his course and began to move slowly towards a lone male Thompson's gazelle. The rest of the pack followed. When the dogs were about two hundred yards from their quarry the gazelle began to run. At first he bounded along in the strange stiff-legged gait known as stotting, but as the dogs started to race towards him, he abandoned that pace for a fast gallop, streaking away across the dry grass. Soon two younger males, Swift and Baskerville, had bypassed old Genghis and the distance between hunters and hunted began to decrease.

For a mile we drove along behind the chase, keeping pace with the dogs that were strung out behind. Finally the lead dogs caught up with the gazelle, Swift running parallel on one side and Baskerville on the other. As the quarry veered away from Swift so Baskerville grabbed its neck. They were running fast and the force of the impact was so great that Baskerville somersaulted right over the gazelle yet without letting go his grip on its neck. He landed with a thud on his back, the gazelle half on top of him, the dust flying up and, for a moment, obscuring our view. A second later we saw that Swift had also seized hold of the gazelle's neck whilst the rest of the pack was converging on the three struggling animals. As the other dogs started to feed, the lead dogs let go of the neck and joined in: within half a minute of capture the gazelle was dead.

A quarter of an hour later no more than a few bones remained, and the dogs set off once more, trotting back in the direction from which they had come. I knew the pack well, for I had first come across it some two and a half years earlier and since then had met up with it on a number of occasions whilst driving across the plains. One of the pack members was missing now – one of the four adult females whom I knew as Juno. Soon I should find out whether the surprising report was true – whether, in this dry, seemingly inhospitable country Juno had a new litter of tiny pups. My hopes were high, for our informant was an acquaintance of our friend George Dove, whose tented safari lodge was the only other habitation of white people, apart from our own small camp, on the short grass plains.

Genghis, again in the lead, trotted steadily for some three and a half miles across the flat plains and then, ahead of us, we saw another wild dog appear as if from the ground and rush towards the returning hunters. It was Juno, her tail wagging and, I was quick to notice, her teats heavy with milk. Frantically she ran from one dog to another, pushing her nose up at the mouth of each, and uttering high-pitched squeaks. She was begging for food, and every so often one or other of the hunters turned or backed away from the persistent female, opened wide its mouth and with a convulsive heave regurgitated some meat on to the ground. Juno gulped down each offering instantly and then resumed her begging.

Many birds and some mammals feed their young by regurgitating food, but the wild dog, like the wolf, has gone a stage further and will also feed a mother who has remained at the den to guard her pups.

After she had eaten, Juno went to a large hole in the ground. She peered down, whining softly, and then started to move down until only her tail was visible above ground. Then she backed out, followed by no less than eight pups. Never before had I seen such young wild dogs and, like the puppies of most domestic dogs, they looked totally unlike their elders. They were, I think, about three weeks old. The pups moved faster than I would have expected for such young creatures, but had obvious difficulty in remaining upright on their wobbly legs and outsize paws. Their ears were large, like those of adult wild dogs, but still completely crumpled, and their dark faces were lined and wrinkled, and more reminiscent of old age than of youth.

When the pups emerged they had no chance either to suckle or to play for, squeaking loudly, the adult dogs immediately descended on them. As the pups stumbled and wobbled hither and thither, so the grown dogs followed, every few moments pushing their noses under the pups and then, with a flicking movement of their heads, turning the youngsters on to their backs. Then, whilst the adults licked their undersides, the pups lay for a moment with all four paws moving slightly in the air before struggling back to their feet and staggering off once more. Often three or even four of the adults gathered to nose and lick the same small pup, pushing each other out of the way, their squeaks following each other faster and faster until the sound resembled the twittering of birds.

I watched as the dominant female of the pack, Havoc, pushed her nose between the hind legs of a running pup as she followed it so that, for a second or so, the youngster ran along on its front legs before tumbling over. As it got up and stumbled off it was intercepted by the fast male, Baskerville, who nosed at it, his front legs straight out along the ground, his rump up in the air and his tail wagging vigorously. He flipped the pup on to its back again with a quick movement of his dark muzzle, and licked. But it was Black Angel, the dark bitch with only half a tail, who seemed to be the most excited by the pups. Her stump wagged so fast that it seemed it might break off and her calls were frenzied. We watched her running alongside a youngster, twittering loudly into its crumpled ear. A moment later she stopped and licked another pup's face so vigorously that it tumbled over backwards. Then she turned her attentions to a youngster that was running from Lotus: together the two females bounded after it, their necks stretched forward. As Black Angel's nose touched the fleeing pup it lost its balance and toppled, head over heels, down into the den.

Next Black Angel ran up to three males who were licking

another pup and, pushing in, tried to keep them away with the side of her body. The males merely moved round and continued to lick the youngster, and Black Angel then made a frantic attempt to keep the pup to herself by placing both her front paws right over it.

Just then I saw Havoc was no longer chasing and licking pups, but had picked one up and was carrying it towards the den. She did not carry it by the scruff of its neck but held as much of its body in her mouth as she could manage. A few moments later she put it down the dark burrow, trotted over to another pup and carried that one to the den as well. Then she took a third. As she was approaching the fourth she was joined by Black Angel who, almost as though she wished to make sure that no pup was dropped, helped Havoc to carry it. Black Angel held on to the youngster's neck whilst Havoc had it by its back. They got it safely into the den and then, together, rounded up and returned the remaining four pups. Then they lay nearby and gradually the entire pack settled down to rest.

By this time it was nearly dark and we had to drive away and leave the wild dogs. I was glad, as I headed for camp, that Juno's den was not too far away from the track leading to Lake Legaja – two years earlier I had found a den so far from any obvious landmark that I had had to drive across the open plains each day on compass bearings.

The discovery of the Genghis pack with pups marked the culmination of five months of searching, and came at a time when I had already given up all hope of finding a den. For a number of years I had watched wild dogs, when time and circumstances permitted, each time that I came across a pack of the nomadic hunters. And, for three years running, I had found dens with wild dog pups in the vicinity of Lake Legaja between January and April. Accordingly, I had arrived at the lake that February, with Jane and her mother and Grub, and the two students Jack and Roger, with high hopes of finding a den.

With Jack and Roger I set out from our camp each day to search for dogs, full of optimism. But as the weeks went by, our hopes gradually ebbed. Day in and day out our three cars had zig-zagged over the vastness of the short grass plains until, ultimately, we were searching an area of five hundred square miles, with occasional forays even farther. During those days we sometimes met up with roaming packs consisting of adults or adults with partly grown cubs, and whenever we did so we stayed with them as long as we could. We worked in relays, relieving each other at approximately ten o'clock in the morning and four o'clock in the evening, until one of us either lost the pack during a moonless night, or failed to find either pack or car if the dogs had hunted over a large distance.

We learned a great deal during such encounters, but we found no den with small pups and that, I knew, was essential for the type of relatively long-term study I wanted to make on the relationships between different individuals in a pack. Only when their pups are young do the nomadic adults stay put for a while in one area of their vast range: once the youngsters are old enough to travel, the home den is left and the pups start out on the nomadic wanderings that they will enjoy for the greater part of their lives.

We were not alone in our searching, for our friend George Dove helped as well. He asked all his drivers to keep a lookout for packs of dogs as they showed tourists around: if a pack was sighted George drove over to our camp immediately to let us know. So an even larger area of country was covered, and we were able to gather information on more of the roaming packs than would otherwise have been possible.

During May and June our hopes rose for we saw mating in two different packs, but both vanished from the area before the pups were due, and search as we would we could find no trace of them. In July, when I had given up all hope, I had to go to Europe for ten days – when I returned, at the beginning of August, George greeted me with the news that his friend had seen a pack of wild dogs with a very pregnant female. And so we had found Juno, with her eight pups, just as I had given up all hope of getting the sort of information I so desperately needed.

The Genghis pack stayed in the area for another six weeks, and during this time Jeff (my new student helper) and I, by keeping an almost constant watch on the pack, were able to learn a great deal about wild dog behaviour.

We soon found that the wild dog pack, like a pack of wolves, has two separate hierarchies, the male and the female. The male hierarchy was much harder to sort out – indeed, we never did, completely. But we knew that Genghis, the old pack leader, and Swift were the two highest ranking males, and all the others came below, Baskerville and Hadis, Rasputin and Ripper, Rinogo and Yellow Peril. Much more fascinating was the relationship between the four adult females, Havoc, Black Angel, Lotus and Juno, the mother.

Juno was by far the most submissive. Whenever Havoc or Black Angel approached her she squeaked, drawing the corners of her lips back into a grin of appeasement, lowering her head and crouching to the ground with her tail wagging faster and faster. Often, too, she presented the side of her neck to the approaching female, a ritual form of submission which I shall discuss later. But no matter how cringing and submissive her behaviour, both Havoc and Black Angel constantly found reason to threaten her and nip her proffered neck in token punishment. And, although Juno was a little less scared of Lotus, she was quite definitely below her in social status.

In particular, Juno had to exercise care when approaching her own offspring. Indeed, on the second day of my observations, I began to fear that the pups might die of starvation, for each time the mother approached the den and the youngsters tried to suckle, the dominant Havoc, sometimes assisted by Black Angel, chased Juno away. I had the impression that Havoc was jealous of the pups' obvious preference for Juno. Each time one of the youngsters tried to follow its mother as she moved away from Lotus's threats, the dominant female picked it up and returned it to the den. However, the following day we saw all the pups suckling.

It seems possible that, through sheer coincidence, we had found the den the very day that the pups came above ground for the first time. Certainly I never again saw the adults greeting the pups in quite so frenzied a manner. Nor, subsequently, did I see Havoc preventing Juno from suckling her young, although this may have been, in part, due to the fact that for the next week of our observations the mother stood in the entrance of her den to nurse her youngsters. This meant that only her head and shoulders were visible above the ground and, since the pups were quite out of sight during the feeds, it is possible that Havoc failed to notice what was going on. On other occasions, particularly when Juno tried to carry a pup, Havoc was quick to rush over and attack the mother.

It always amused me to see Havoc marking the area around the den with her urine. This sort of marking often indicates ownership of territory, and I could not help wondering whether, perhaps, she was trying to stake out some kind of claim over the pups. Probably, though, she was simply asserting the prerogative of the dominant female, for I never saw the other females mark around the den. Perhaps her message read: 'Here is the den of the Genghis pack. Here I, Havoc, am the dominant female. Go no farther!'

Black Angel, who was second to Havoc in the female pecking order, was completely fascinated by the pups although, being less dominant, she was slightly more cautious than Havoc in her dealings with them. Thus when she went over to another dog who was licking a pup it always looked quite by chance that she moved between the two, and quite accidental that she pushed the other away, very gently, with the side of her body. But it happened so often that I knew her manœuvres were purposeful. And frequently if another adult tried to join in when Black Angel herself was licking a pup, she would promptly lie right on top of the youngster, almost hiding it from sight – and, I felt sure, half suffocating it too. Usually the other adult then wandered away in search of a more accessible pup to lick.

Black Angel was also fascinating to watch when she interacted with the other females. Constantly she ingratiated herself with Havoc. If the dominant female moved in her direction, Black Angel hurried towards her, wagging her tail and squeaking, her ears back and her hind legs slightly crouched in submission. When they met

Black Angel licked or nibbled Havoc's mouth and often repeatedly rubbed her chin over the dominant female's nose and head. Sometimes Havoc only needed to move a few yards to a new resting place for Black Angel to run over and make these submissive gestures.

One day, soon after we had found the den, I saw Black Angel take a few steps towards Juno, who was licking one of her own pups, and then hurry over to Havoc who was standing some way away, and briefly rub her chin over the dominant female's head. This done, she ran back and began to nip Juno in her neck. To my surprise, Havoc followed and began, in turn, to bite Black Angel. At first I was puzzled, but this sequence of events repeated itself so often during the days that followed that I soon realised what was going on. It seemed that Black Angel, just prior to attacking either Juno or the other female, Lotus, either attempted to get the dominant female's permission for such an attack, or at least tried to ensure that Havoc would remain a non-involved observer. If, in fact, this was her intention, she was normally unsuccessful, for nearly always when Black Angel bit another female, Havoc, as though to keep the order, promptly bit Black Angel. Usually, though, Havoc's bite was quite obviously gentle: sometimes, indeed, it was immediately followed by gentle, nibbling, grooming movements as though the dominant dog, after reprimanding Black Angel, wanted to assure the other of her continued friendship.

It always seemed to me that Black Angel felt that her own high-ranking social position depended in part on maintaining friendly relations with Havoc, in part on ensuring that the same friendly relations did not develop between Havoc and either of the other females. Often I saw Black Angel go to great lengths to keep Juno or Lotus away from Havoc. Thus, if Lotus approached Havoc, Black Angel usually hurried over and either placed herself between the two dogs or pushed Lotus away with the side of her body. In either case she sometimes bit Lotus quickly as well – for which breach of etiquette she was usually, as I have said, reprimanded by Havoc.

Sometimes it seemed that the anticipation of her own punishment on such occasions set up a conflict within Black Angel. She would make a fast movement towards Lotus, as if to bite her neck, then turn quickly and make a fast movement towards Havoc as if to chin-rub, but without actually touching either. After she had repeated these two intention-movements several times, Lotus usually moved off and Black Angel escaped Havoc's nip.

Lotus, although she was dominant to Juno, spent less time than the other females with the pups, thus avoiding the centre of activity. When she did approach the youngsters, Black Angel often hurried up and pushed herself between and, if Havoc was not too close, she usually nipped Lotus's neck as well.

Jeff and I usually spent whole days watching at the den, and soon the approach of either of our cars merely caused some of the dogs to raise their heads briefly whilst others did not even pay us that much attention.

The adults went out hunting in the evening, during moonlit nights or very early in the morning. During the day they rested near the pups. Let me describe a typical day when the pups were about weeks old.

During the first part of the morning the adults, for the most part, lay around in small groups, resting or sleeping. The pups played together close to the den and, every so often, one or more of the adults wandered over to nose and lick them. At about 10.30 a.m. the pups disappeared into the coolness of their den and soon the adults wandered off, in ones and twos, to lie in nearby burrows. For a while Juno went down with her pups, but soon she reappeared, shook herself, and went off to lie in another den nearby. A little later, and somewhat to my surprise, the adult male, Swift, left his resting place and went down into the den with the pups. He was there for well over an hour, and soon after he had left Black Angel joined the pups for a while.

At about half past four the sun disappeared behind some heavy cloud, and soon after this three of the adults appeared from their dens, almost at the same time. Together they headed for the pup den, running along side by side, twittering and licking and nibbling at each other's faces. One after the other they put their faces down the hole and whined, their ears pricked forward and their tails wagging. Suddenly Havoc rushed past them and right down into the den. I heard squeaking and twittering from the depths of the earth and then Havoc backed out, followed almost immediately by the eight pups.

For several minutes there was confusion as the adults greeted the pups and each other and, one by one, all the other members of the pack appeared and joined in. But soon things calmed down, and the adult dogs lay down again and rested, lying in the open under the clouded sky.

The pups began to play again. Still unsteady on their legs, they wobbled about, biting and pulling at each other's crumpled ears, tumbling over each other as they wrestled. Black Angel lay close by them and every so often one of the pups crawled over her legs or her tail and was promptly tipped on to its back and subjected to a vigorous licking.

Just as the sun was setting old Genghis rose to his feet and yawned as he stretched himself. He trotted over to where Havoc, Swift and Baskerville lay together. At his approach they jumped up and all four began nosing and licking each other's lips, their tails up and wagging, their squeaks gradually changing to frenzied twittering. In a moment all the adult dogs had joined them and soon the pack was swirling round and round in the greeting ceremony. Amidst the confusion of legs and tails and lean lithe bodies I caught a glimpse of Havoc and Swift, their wide open mouths touching, their tongues curled back in their mouths; a momentary flash of old Yellow Peril piddling all over his toes in excitement; a sudden picture of Juno, her forelimbs flat on the ground and her rump up in the air as she twisted round to lick Genghis on the lips. And then, as suddenly as it had begun, the wild flurry of activity subsided and the pack started to trot away from the den on its evening hunt. This ceremony, which nearly always takes place before a pack sets off hunting, can best be compared to our own 'Good morning' – a married couple will often kiss each other when they wake even if they have been sleeping side by side, and in Germany all members of a family shake hands every morning and evening. Most of the gestures which occur in the ceremony appear to have been derived from begging behaviour – there is the same nosing and licking of the lips. And, in this context, there seems to be little or no difference between the behaviour of a high-ranking and a low-ranking individual. It seems, most perfectly, to express the unity of the pack in hunting: 'I submerge my identity,' the twittering cries seem to say. 'I will do my share of the hunting, I will share in the feeding. Let us go! Let us go!'

The mother, Juno, accompanied the pack for two hundred yards or so, but then returned to guard her pups. The youngsters themselves made no attempt to follow but stayed playing close to the den entrance. I have watched three other packs with pups and, in each case, the mother remained at the den whilst the rest of the pack went in search of food. In each of these packs there were eight or more adult wild dogs so that the temporary loss of one member of the hunting unit did not endanger a pack's chances of successful food-getting. Certainly the Genghis pack, with twelve adults, could well afford to leave Juno behind.

The old male Genghis was normally undisputed leader when the pack was on the move, for it was he who determined, on nearly every occasion, when and where the dogs should go. Once when the pack set off Genghis was in fourth place. After the dogs had trotted for about a mile Genghis veered to the right whilst the front dogs continued straight on. But within thirty seconds the apparent leaders had also changed course to the right so that Genghis, although he remained in fourth place, had clearly initiated the change of direction for the pack as a whole.

On this occasion Genghis was walking some ten yards in front of the others, as he usually did. The rest of the pack was strung out behind him, mostly in single file. I drove slowly parallel with the dogs, taking notes on individuals as they changed their order of progression and timing the pack's trotting speed – approximately seven miles per hour. Often the dogs paused, singly or in groups, as they investigated a hole in the ground or sniffed around a clump of high vegetation. Once Lotus stopped to eat the small brown pupae cases that protruded from some old wildebeest horns. These short tubes are left adhering to the horn when a species of moth, related to the clothes moth, burrows out after pupating. They are very common, but this was the first time I had seen a wild dog eating them – or, for that matter, eating insects of any species.

I did not see many animals as we crossed the dry, barren country, and the wild dogs did not start their first chase until we were about five miles from the den. Then three Grant's gazelles became visible in the dusk, and soon the dogs were after one of them. On the smooth plains it was not difficult for me to keep up with the hunt. Genghis, although he had been the first dog to run, was soon passed by Swift, Havoc and Baskerville. After five minutes Black Angel ran into second place, and I stayed level with her. She was running no slower than Swift, but I avoided going too close to him as I did not wish to add to the fright of the gazelle and thus influence the outcome of the hunt.

For the next three and a half miles my speedometer did not leave the thirty miles per hour mark: the dogs kept up that speed, at least, over the whole distance. Occasionally one of them made some headway for a short distance, possibly touching thirty-five miles per hour or even faster.

After the first three miles of the chase Swift was still in the lead, and Black Angel was still running second, but when the gazelle suddenly began to circle round I saw Baskerville quickly swerve and cut the corner so that soon he was leading the pack. And then, half a mile later, when Baskerville and Swift were only yards behind their quarry, having slowly gained ground all the way, they suddenly appeared to give up. One by one the dogs stopped running and the pack, which had become very spread out during the hunt, gradually reunited. The gazelle, still running, soon vanished into the fast falling night.

That chase, in fact, was one of the longest I have observed: usually, if a pack of wild dogs fails to bring down its quarry within two and a half to three miles, it abandons the attempt and, after resting for a while, sets out after a different prey. This, of course, disproves one of the old stories about wild dogs, namely, that once a quarry has been selected it is doomed and will be relentlessly chased to exhaustion and then captured.

Over the years my assistants and I have watched wild dogs hunting on numerous occasions, and of the ninety-one actual chases involved, thirty-nine were successful. Despite widely held beliefs to the contrary, a pack of dogs trotting across the plains does not necessarily arouse immediate panic in the herds of prey animals. When the plains are black from one horizon to the other with the migratory herds of wildebeests and zebras, those nearest to the dogs usually trot or canter quite slowly out of the way and then turn to watch as the hunters continue on their way. It is usually only after a pack has made a few unsuccessful chases in succession, or when it has been hunting for a long time in one area, that prey animals panic at their approach. Once the dogs begin to run, however, rather than just walk or trot, all prey animals within a few hundred yards usually begin to hasten away.

It is probably for this reason that wild dogs normally approach a selected individual or herd very slowly, walking with their heads held low and parallel with the ground, and adopting a slight crouch in their gait. In this way the hunters can sometimes get to within fifty yards or so of a herd of zebras or wildebeests before the intended prey begins to run: a herd of gazelles, however, usually takes off when the dogs are still at least one hundred yards away.

Once the quarry begins to run the wild dogs usually begin the chase in earnest. And, from this point, the hunt may develop in a number of different ways. Sometimes, particularly when the dogs are approaching a small herd, it appears that the prey is selected before the chase begins, often by the leader of the pack. When he starts to run the other dogs follow suit, and all dogs pursue the same prey until it is either caught or manages to escape. At other times, usually when the wild dogs are hunting large herds, the pack makes a short chase towards a group of animals and then stands or walks slowly whilst watching intently as the herd starts to run. Then the dogs may chase one animal, or they may trot off and repeat the same tactics with another herd. Yet again, after a herd has begun to move, the dogs of a pack may split up and several different chases may go on at the same time. Very occasionally two such chases will be successful, particularly during the birth season of the wildebeests when the dogs go for the calves. More usually, however, the different chases culminate in one. It is as though each dog keeps alert as to the possibilities of success in the chases other than his own, and will leave his quarry if he notices that another dog or group of dogs is having better luck.

It is the last two techniques which are the most fascinating. What are the dogs looking for when they stand watching a herd run past them? And why do the individuals of a normally closely united pack sometimes separate in seeming disorder? The answers are, I believe, the same: both techniques enable the wild dogs to select an individual from the herd that is, in some way, weaker and slower than his fellows. It has been maintained that wild dogs have no need to pick out weaker creatures because their speed and stamina are such that they can run down the swiftest of prey: however, since a certain proportion of their selected quarries escape, this seems unlikely. I am convinced, although proof is difficult to acquire, that when the dogs set a herd running it is so that they can more easily pick out an animal less fit than his fellows. And I should point out here that this animal does not have to be actually lame or sick: wild dogs are undoubtedly able to pick out far more subtle signs than we can which indicate that a beast is off colour and, therefore, a suitable victim. And, if the review of running animals fails to pinpoint an obvious candidate, then a number of separate chases after the scattering herd are more likely to find a laggard than one combined effort.

During the actual chase wild dogs placed behind the leaders will cut corners when the prey changes direction and so gain positions farther forward in the pack, or actually in the lead. This is particularly noticeable when the dogs chase a Thompson's gazelle, for this creature usually zigzags across the plains when hunted, or describes a very large circle, so that, by the end of the chase, several different dogs may have led the pack simply by cutting the corners. It is this which presumably gave rise to the once firmly held belief that the wild dogs hunt in relays, fresh dogs which have been running slowly in the rear racing to the forefront of the pack when the leaders tire. In fact, during all the chases I have watched, the original lead dog or dogs normally kept up the same speed when those behind cut corners, and often ran even faster and thus regained leading positions once more. It is true that some dogs travel slowly in the rear – Yellow Peril was often half a mile or more behind the leaders when a kill was made. But I never saw him rush forward to take the lead.

The wild dog hunts a large variety of prey animals including warthogs, Thompson's and Grant's gazelles, wildebeests and zebras, all of which I have watched them chase and kill. Until recently it was thought that wild dogs would not tackle zebra; for one thing the zebra is a powerful and heavy animal; for another, the stallions attack in defence of their herds; for a third, zebras seldom run until a pack of wild dogs is very close. Nevertheless, in some areas zebra are frequently preyed upon by wild dogs, and one year, during the migration, I watched the Genghis pack hunting them on many occasions. That was before Juno had her pups.

One such hunt, in particular, is worthy of mention. Genghis, in the lead as usual, began to move slowly, in the stalking attitude, towards a herd of about twenty zebras, amongst which was a mare with a small foal. As the dogs got closer they seemed to move more and more slowly, giving the impression that they were not hunting at all. In this way they actually got to within twenty yards before the zebras, which had been standing watching the dogs, turned and trotted away. This was the signal for the chase to begin in earnest, and the dogs ran in pursuit. The zebras began to canter, closing their ranks until they were running in a tightly packed group. As they ran across the plains other small zebra groups in the area joined them, so that soon the zebras numbered well over fifty. They did not run fast, but rather seemed to be keeping pace with the slowest of the group – in this case probably the foal.

At the start of the chase Genghis maintained his leading position, with Swift and Baskerville close behind. Suddenly the old male put on a spurt, but as he drew close to the mare with the foal, who was at the back of the herd, so one of the stallions turned and lunged at him, ears back and teeth bared. Quickly Genghis swerved away, so that Swift was in the lead. After a few moments Swift closed the gap between himself and the mare; again the stallion turned and attacked in defence of his family. This happened a number of times; in each instance the lead dog avoided the stallion, leaving another dog to make the next rush forward.

Ultimately, in the confusion, the mare with her foal, and also a yearling which, from its stripe pattern, I guessed was the mare's elder offspring, became separated from the herd. Immediately the wild dog pack surrounded the three whilst the other zebras soon vanished over a slight rise.

Left on her own, the mare stopped, the foal and yearling close beside her. As Swift, Baskerville and two others moved towards the zebras the mother took a step forward and bit towards Swift. As she did so, three dogs moved rapidly up behind the foal, but they were headed off by the yearling which lunged forward in the foal's defence. Time and again the dogs tried to grab the foal, but each time they were turned back by mother or yearling, neither of which, no matter what happened, took more than a few steps in any direction. The foal was able to maintain its position, pressed close to its mother's side, and none of the three were, for a moment, separated.

It was a tense situation and I wondered, somewhat desperately, how long it would continue. As the minutes went by the dogs became bolder, closing in from all directions, some squeaking as if to raise their morale or to egg each other on to make a grab at the prey. All at once Swift made a lunge towards the mare's head, his jaws snapping as he grabbed for her upper lip. This is a pattern which frequently occurs when wild dogs capture an adult zebra – one dog seizes the upper lip and pulls hard whilst the rest of the pack disembowels the prey. Domestic hunting dogs, which have brought a quarry to bay, sometimes show the same pattern. Such a hold probably acts in the same way as a twitch – a rope which, when twisted around the upper lip of a horse, enables the handler to hold his animal still for medical attention, and so on. The wild dog which holds a zebra in this way normally maintains his grip until the prey is all but dead before he joins the rest of the pack in feeding.

On this occasion the zebra mare jerked her head up and Swift's teeth closed with a snap in the air. But by now his blood was up and again and again he leapt forward. The end seemed inevitable – and the end is always much worse to watch when the prey has bravely defended itself or its young one. But suddenly I felt the ground vibrating and, looking around, I saw, to my amazement, ten zebras fast approaching. A moment later this herd closed its ranks around the mother and her two offspring and then, wheeling around, the whole closely packed group galloped off in the direction from which the ten had come. The dogs chased them for fifty yards or so but were unable to penetrate the herd and soon gave up. That was the only time I saw zebras return to others of their kind surrounded by a wild dog pack, thus enabling the victims to escape.

It is, of course, for their method of killing, by disembowelling living prey, that the dogs have been most hated, and for which, in many areas including some National Parks and Game Reserves, they have been almost exterminated. Of course it is a sickening sight to see a wild dog pack tearing into a victim's groin. But, at the same time, it is questionable whether the animal feels as much pain as one imagines. The accounts of people who have been mauled by lions – Dr Livingstone is a good example – and those of men who have been badly wounded in war, often reveal that, due to physical and sometimes mental shock, deep ripping wounds are not felt by the victims until some time later. It often seems, at the time, that a pack of wild dogs takes an eternity to dispatch its unfortunate victim, but of the thirty-nine kills that we have timed, all but one took no more than five minutes before death occurred, and frequently the prey died in less than two minutes. In the one remaining instance a yearling wildebeest, caught by a pack of four dogs, took seventeen minutes to die. There are accounts of long drawn-out kills by large packs of wild dogs when the prey took twenty-five minutes or so before it was released from its suffering by death: in those cases I know of this has been due entirely to the human observers – by driving too close they disturbed the hunters so that some members of the pack hung back apprehensively and were not able to do their share in the killing. There are, indeed, some published photographs which show this only too clearly.

And so, viewed dispassionately, the wild dogs, like domestic hunting dogs, wolves and hyenas, are quick and efficient killers. They attack their prey where the skin is thinnest and thus quickly reach the internal organs and dispatch the victim. Throttling, the system used most often by the cats to kill their prey, is less gory and is, therefore, considered to be a 'kinder' way of killing: sometimes, though, it takes ten minutes for a victim to die and may, for all we know, cause more suffering to the animal concerned. I should mention too that, provided the prey is small enough or rendered harmless by means of a broken back, lions, leopards and cheetas will also eat their victims alive and, on those occasions, the suffering of the prey concerned is very long drawn-out and horrible to watch, for the cats, unlike the dogs, take their time when feeding.

One morning, as I was driving towards Juno's den, I came across Genghis and the other adults also heading in that direction. Their bellies looked full and their heads and necks were dark with dry blood. Obviously they had recently made a kill. Slowly I followed them and soon I became part of the pack, for old Yellow Peril and Black Angel dropped behind and trotted along after the Land-Rover. As we approached the den I could see Juno sitting with her offspring near the den entrance. It was a week since I had first seen the pups: now they were steady on their feet and some of the wrinkles on their faces and ears had straightened out. As we drew near, Black Angel and Havoc ran on ahead. Juno, as usual, hurried to meet them, cringing and begging for food, but the two females dodged past her and made for the den. I was surprised to see that all the other adults avoided Juno too, for normally many of them responded to her squeaks and licks by regurgitating meat. And then I suddenly noticed that the hunters, one after the other, were regurgitating to the pups.

Like most carnivores wild dog youngsters start eating solid food when they are still very young – probably at about a month old.

As the days went by the pups became more and more eager to get a good share of the food brought back by the hunting adults. Two days after I first saw the pups eating meat, Demon, one of the male pups, pushed his head right in between Lotus's teeth as she opened wide her mouth ready to regurgitate the meat. Frenziedly Demon reached to the food as it came and gulped it down before his siblings could grab some of it. After this it was quite common to see the pups pushing their heads thus into the open mouths of the adults.

At first it seemed that the pups had difficulty in tearing up pieces of meat that were too large for them to swallow whole. Then tremendous tugs-of-war took place. I often saw Havoc join in such a battle – sometimes she got the piece of meat and ate it, in which case she usually regurgitated it again for the pups later. Once, when two pups were each tugging at one end of a piece of skin, Demon raced across to join them but, failing to notice the den, fell straight in and vanished from sight.

Once the pups had started to feed on meat, Juno often seemed to have difficulty in getting enough food for herself: the other adults obviously preferred to feed the pups. Whenever the pack returned from a hunt Juno still rushed to meet them, frantically pushing her nose to the mouth of each dog, squeaking and twittering and wagging her tail. Occasionally one of the hunters responded – most often Genghis – but for the most part the others quickly jumped past the mother and rushed to feed the pups. This meant that in order to get enough food for herself, Juno had to jump in amongst her pups and take a share from them. And, as might be expected, this meant that she was attacked even more frequently by Havoc and Black Angel.

It would be wrong to suppose that Juno was unsuccessful in her begging simply because of her very low-ranking position, for, in another wild dog pack I watched, the mother was a very dominant dog yet she too was usually unsuccessful when she begged from the other adults. Like Juno, that mother remained at the den to guard her pups, and like Juno she jumped in amongst her youngsters in order to get a share of the food brought back by the returning hunters. It was obvious, however, that she managed to get more than enough for her own needs, for afterwards she invariably regurgitated part of her meal to her pups.

That mother gave birth when the plains were green and the migration was all around. The wild dogs had food in abundance: the other seven adults of the pack, all males, hunted twice a day, early in the morning and in the evening, and they were almost always successful. And they needed to be, as the pups grew older, for the mother had produced an extraordinarily large litter. Indeed, it took me several days of counting and recounting before I was quite sure of the exact number of pups at the den – no fewer than sixteen. One wild dog bitch in captivity gave birth to nineteen pups, but three were still-born: perhaps this wild female had also produced a few dead babies.

As the wild dog mother has only twelve or fourteen teats the sixteen pups could not all nurse at the same time, but surprisingly I saw no fights for a feeding place when the youngsters suckled. Instead, a few of the pups remained on the outskirts, calmly waiting to take their turn when the first ones had had enough and moved away. It often seemed, though, that the last comers could not have drunk their fill when the mother moved off, for the suckling bouts lasted no longer than those of mothers with far fewer pups – an average of two and a half to three minutes.

A wild dog mother usually nurses her young in a standing position, and often the pups stand upright to reach her nipples. Sometimes they make kneading movements with their front paws against the mother's belly, sometimes they maintain balance by placing their paws on the back – or head – of a sibling. Often a mother terminates a suckling bout by merely sidestepping over the pups and moving away. But the mother of the sixteen, surrounded by such a horde of pups, usually jumped right over them at the end of a feed, leaving a split second scene of two rows of heads with pouted mouths, some still making sucking movements. Often one or two pups lost their balance as the teats to which they had been clinging suddenly disappeared.

That mother's nursing problems revolved entirely around the enormous size of her litter: Juno's nursing problems were of quite a different nature. The first difficulty, as we have seen, was when Havoc, the dominant female, chased Juno away from her own pups. Subsequently Juno suckled her pups down in the entrance of the den. However, the youngsters were growing fast, and possibly it got rather crowded down there at feeding time. Then Juno nursed them out in the open again. For two days all was well – it seemed that Havoc had accepted the fact that Juno, as a mother, had certain duties that had to be performed. And so Juno would go over to the den, call her youngsters with soft whining sounds similar to those of a domestic dog, and suckle them in peace. But then her troubles started all over again, for Lotus and Black Angel began to join the pups at Juno's teats. At first we only saw the dominant females licking at a nipple which a pup had abandoned, but it was not long before they were, so far as we could tell, suckling like two youngsters. As one of the high-ranking females approached, Juno, grinning and wagging her tail, often flopped submissively to the ground and, as pups and adults attached themselves to her nipples, she was forced to maintain this position for the duration of the feed. On two occasions I saw Baskerville suckling from Juno, and Lotus, the other female, sometimes tried. Once she got there before Havoc, and when she saw the dominant female hurrying along to join in, Lotus panicked and, taking the direct route of escape, pushed her way under Juno's tummy, scattering the suckling pups in all directions.

I wondered, for a time, whether these adults behaved in this strange way because of the hot dry weather, for no rain fell during the first few weeks of the pups' lives, and the nearest water that we knew of – a very brackish pool – was all of ten miles away. Of course, that is not so far for a pack of wild dogs, but certainly they never drank on all the occasions we followed them and, moreover, another scientist who watched a pack with pups for a two-month period saw no drinking in the adults either. It is just possible that Juno went off to drink by herself for once, when we arrived at the den in the morning, she was not there and she appeared, one and a half hours later, trotting over the horizon from the direction of the brackish water. Certainly it would seem that a nursing mother, more than any other adult, would need to replenish her liquid intake, and when it did rain a little Juno was more energetic than the others in licking the moisture from her own hair and the coats of the other adults and pups. But the rain did not stop Havoc and Black Angel from suckling: possibly it had, by then, become a habit.

It was not only Juno who suffered from the adults' desire to nurse: often when Havoc and Black Angel suckled side by side from the lying mother they took up so much room that some of the youngsters were unable to get to the teats at all and sometimes the two females were almost hidden from sight by pups that clambered all over them as they tried to squirm in to nurse. Once I saw two pups together pulling at Black Angel's stump of a tail: probably they were simply having a game, but it looked exactly as if they were trying to pull her out of the way.

Tail-pulling is, in fact, a very frequently observed game in wild dog pups, and Juno's youngsters pulled each other's tails and ears quite vigorously during many of their play sessions. Often, too, they would pull the tails of adults lying near. Juno stood no nonsense from her pups. Once I watched as the pup Demon crept up behind her (in exactly the posture used by an adult stalking its prey) and pounced on his mother's tail. Juno turned quickly and made a slight lunge towards him, her mouth wide open. Undeterred, Demon pounced again and began to pull her tail, uttering little growls. Again Juno threatened and again the pup ignored her. The fifth time Demon pulled her tail, Juno bit his nose. As the mother laid her head on her paws again, Demon sat quite still staring at her tail. He glanced towards Juno's face, gazed back at the tail and then, after a final look at her face, got up and wandered away.

Black Angel's reaction was rather different. One day, as she lay asleep, the pup Sprite stalked up behind her, pounced on her tail and bit it. Black Angel looked up quickly, as though startled. She made no other movement, however, except to kick backwards with one leg. She caught the pup in the chest and Sprite, a tiny ball of fur, rolled head over heels backwards for a couple of yards. Slowly Sprite got up, stared at the motionless Black Angel, and moved away. Meanwhile Demon had started to stalk up to Black Angel. Suddenly he pounced, caught her tail in his mouth and pulled hard. This time Black Angel did not even look up. Again she kicked out: Demon flew through the air in a sitting position, a quizzical look on his face, and landed in a cloud of dust. For a moment he sat there, staring towards Black Angel who lay motionless on her side. Then he got up, moved towards Sprite, and the two of them crept up on Black Angel side by side, inching forward on their tummies. They were about one foot from her tail when she slowly raised her head and stared at them: immediately they flopped down and lay quite motionless. Then Black Angel jumped up, flipped them both over with her muzzle, one after the other, and licked them vigorously as though she wanted to make it quite clear that the whole thing was a game.

I was rather surprised to find that, throughout the seven weeks that we watched the den, we never saw the adults actually playing with the pups. Two years previously, when I watched the Genghis pack for a short while, they had been very playful, particularly after feeding and when they got up from a long rest. Their games were often wild: two would stand upright with their front paws on each other's shoulders as they bit each other's necks and then they would chase each other round and round, leaping over each other as they ran. Swift and Baskerville, in particular, after jumping over a companion, often turned a complete somersault as they landed on the ground. At that time, too, they had a den with pups, offspring of a female who had since left the pack or died, but they were surrounded by the migratory herds, the grass was green, pools of water were everywhere. It was these things, perhaps, which induced their playful moods. One pack that was watched for quite a while in the Ngorongoro Crater where there is always plenty of food and water, also played frequently.

However, there was one occasion, when Juno's pups were young, which evoked playful behaviour in some of the adults. It was one of the mornings when Jane, with our son Grublin in the car, had joined me near the den to watch the wild dogs. As usual, at that time of year, a strong wind was blowing across the plains. For a while Grublin was very quiet – he was sitting in Jane's car painting a picture. His work of art finished, he stood up and held it out of the window for me to see. It was on a large piece of paper, folded in half, and the wind instantly snatched it out of his hand. As the paper was blown over and over across the ground the dogs rushed out of the way, and several of the adults gave the gruff threat bark. But then they seemed to realise that the strange object was not dangerous, and Black Angel, Swift and Baskerville set off in pursuit. Black Angel darted in front of the paper, but backed frenziedly and leapt aside as a sudden gust of wind blew it straight at her face. Then she and the others were after the paper once more. Finally Swift grabbed it in his mouth, tossed it into the air, and made a tremendous leap to catch it again as the wind seized it away. Ultimately Black Angel came trotting back with a few soggy shreds still in her mouth. She lay down, chewed for a while, and then spat out the pulp.

The wind that sweeps daily across the Serengeti Plains is an integral part of the lives of all the creatures that live there. Sometimes, in our camp, we cursed the wind for it blew dust in our eyes and our food, whirled precious papers away across the grass, and slowly and relentlessly destroyed our tents. The constant flap-flapping of some of the older canvas was fraying to the nerves, for the replacement of a canvas fly sheet costs close on £100.

One morning, when Jeff was watching the wild dogs, I sat typing my notes in camp. Gradually the wind began to blow more and more furiously until I could scarcely hear my voice on the tape-recorder. I looked outside and saw that half the sky had been covered by thick black clouds whilst, to the west of the lake, the soda dust was being blown at least a mile up into the air by the gale. The sun was still shining, and the dust was snow-white against the blackness of the sky. It was a scene of awe-inspiring grandeur, but I had little time to gaze at the beauty for I realised that we were in imminent danger of losing our tents. Jane and I and the two

African staff spent a hectic fifteen minutes rushing round tightening guy ropes and closing zips – and, as it was, the wind ripped through the canvas along the side of one tent as though it had been tissue-paper. However, compared to some people I know, we were lucky. These were the clients of a large safari party, travelling round the National Parks. They were snugly asleep one night when a sudden gale swept through their camp and tore two of the large tents right out of the ground, leaving the terrified clients clinging to their bedclothes and exposed to the fury of the elements. The tents, torn to shreds, were recovered the next day from the branches of nearby trees. Our own tents have sewn-in groundsheets and I have always wondered whether a really strong gale might not take us and our beds flying away with the tent.

Even the wild dogs, it seems, are sometimes bothered by the wind. The first time I noticed this was early one morning, when the sun had only just risen and the wind that shrieked over the flat plains was cold from the night air. The pups were snugly down in their den, and the adult dogs were lying separately except for Swift, Havoc and Black Angel, who lay together, the sides of their bodies touching. Gradually the strength of the wind increased, and presently those dogs who had been lying alone joined the group of three. Each dog, as it arrived, lay on the leeward side of the dogs already there, protected by the bodies of their companions from the full force of the gale. Yellow Peril, the old dog with half a tail, was last to join the others and he, too, lay on the leeward side. Within the next few minutes, however, Havoc, who was lying directly in the wind, got up and lay in the shelter of Yellow Peril. Soon Black Angel, then Swift followed suit, and then the other dogs, one after the other, until Yellow Peril was acting as the windbreak. Maybe he was hardier than the others for, during the next half-hour, there was peace. Then he too moved to the leeward side of the pack. Within the next six minutes all the dogs moved, one after the other, until Yellow Peril was once more left to bear the brunt of the wind. This entire sequence happened twice more before the heat of the sun made the strong wind a blessing rather than a curse. Subsequently Jeff and I watched the dogs behave in this way on many cold mornings: always Yellow Peril endured the discomfort for longer than the other dogs.

We never saw the pups resting in a group with the adult dogs, and always, when the wind was cold, they took shelter in the den. Indeed, the dens of young wild dogs, like those of jackal and hyena youngsters, are places in which to hide from all the dangers of the outside world. I saw Juno's pups dart underground in response to many potential dangers – a bird flying low overhead; the close approach of some hyena or other animal; the sudden rapid arrival of an adult of their own pack; and so on. But the incident I shall never forget occurred early one morning. I arrived at the den just as the first light of dawn slowly gave substance to the surrounding country. There was a strip of low-lying black cloud on the horizon, but soon the sun rose out of this, a dark red globe. All at once I noticed that the four pups who had been lying at the den entrance were staring at the rising sun, their ears pricked: after a moment they shot down into the den. But after a few seconds the four heads slowly appeared over the edge of the burrow again, their foreheads wrinkled as if pondering some great mystery. Suddenly all heads jerked back into safety again, only to reappear once more, the large ears first and then, slowly, the wrinkled foreheads. With their eyes at the level of the grass around the den the pups stared and stared towards the sunrise. Why it made such an impression on them that day I cannot guess; if it really was the sun they gazed at. Maybe, though, they were looking at something I could not see. Afterwards, though, Jeff swore that the pups played peek-a-boo with the sun one morning, popping in and out of their den as a whole host of small clouds drifted across the sun, one after the other.

As I was concentrating on the relationship between the twelve adults of the Genghis pack I was not able to make many detailed observations on the behaviour of the pups. I learnt to distinguish one from the other, and gave them names, but I did not watch them enough to learn their different personalities. If they had a ring-leader it was Demon. He always seemed to be the roughest pup, the most persistent in begging from the adults, the only one I ever saw pull the tail of old Genghis. If, during play between two pups, one of them began to squeal and tried to escape from the game, the other pups often stopped playing, rushed over to the scene and, like a gang of Teddy-boys, attacked the squealing pup. During such mobbing scenes any of the pups were likely to be victims, but Demon always seemed to be in the forefront of the attacking gang.

As the pups grew older the adults became less tolerant. Havoc was especially quick to turn when a pup followed her too closely and, with her open mouth around its neck, hold it for a few moments pinned to the ground. This was a typical punishment, and often the pup squealed loudly during the process.

Obstreperous pups were also reprimanded by quick, and usually gentle, bites on their muzzles or necks, at which they normally rolled on to their backs and lay with all four paws in the air until the adults moved away. The pups earned many of their punishments at feeding times, for they would persist in begging even after an adult had twice regurgitated. Once the pup Sprite made herself such a nuisance that Genghis ran at her, biting her neck, and sending her rolling head over heels in the dust. She squealed loudly but did not appear to be hurt.

The urge, in an adult wild dog, to regurgitate to pups appears to be strong. In a pack studied by another scientist in Ngorongoro Crater the only female died when her nine pups were only five weeks old. The male adults of the pack, however, continued to care for the pups, returning to the den day after day to feed them until they were old enough to accompany the pack on its hunting forays.

Some zoologists maintain that regurgitation is a response to the begging behaviour and squeaking of the pups, or of begging adults. But this is not necessarily true, for I have seen adults who have been peacefully resting for some hours suddenly get up, walk towards the playing or resting pups and, without prompting, regurgitate meat on to the ground. In fact, although an adult, after regurgitating once or twice on returning to the den after a hunt, may reprimand a pup that begs for more, it is quite likely, later on, to regurgitate yet again. Often, indeed, a dog may regurgitate hours after returning to the den.

It was late afternoon, when the pups were about six weeks old, that I saw them moving, with the pack, to a new den some forty yards from their original home. I am not at all clear as to how the journey was initiated, for there was a sudden confusion of dogs, rushing about squeaking and greeting each other, and then the whole pack, together with the pups, began to move off across the plain. Perhaps the wild dogs have a special call, signifying 'Follow': but this must remain a guess until detailed sound recordings have been taken and analysed. Anyhow, whatever it was that urged the pups to follow the adults from the den for the first time in their lives, follow they did. Soon the pack arrived at the new den, where there were three burrows quite close together. Whilst the pups eagerly sniffed around their new surroundings, Havoc and Black Angel began rushing from burrow to burrow, sometimes digging briefly in one, only to leave it and hurry to another. It seemed as though they could not decide which one was the more suitable for habitation. Juno did not join the two dominant females in their inspection of the dens: instead she kept trying to pick up one of her pups. But after getting it into her mouth she would lose her grip, so that the youngster fell to the ground. Where she wanted to take it I could not tell, for Black Angel suddenly noticed and rushed over to bite Juno in her neck – the pup, of course, made a quick getaway.

Soon, to add to the general confusion, the eight pups joined Havoc and Black Angel in their investigations. Once Demon followed Black Angel down one of the dens only to emerge hastily, half hidden in a cloud of dust, as the female started digging.

After ten minutes or so it seemed that Havoc and Black Angel had each decided on the most suitable den: but the choice was different. Black Angel grabbed one pup by the skin of its back and, followed by two other pups, vanished into the den of her choosing. Havoc, at the same time, took another pup in her mouth and went into *her* den. Then I saw Juno also trying to pick up a pup, but each time she tried to grab it by the skin of its neck or back it rolled over, presenting its round tummy and kicking and struggling frantically as its mother tried to roll it back again with her nose. Ultimately Juno grabbed its ear between her teeth and began to drag it along the ground, but at this moment Black Angel emerged from her den and instantly raced over and bit Juno in her neck. Again Juno's pup made its escape.

The scene became even more confusing as the two dominant females took pup after pup to their dens: each time the pup in question was left, it immediately ran off to explore again. Three times I saw Havoc marking around the den she had chosen. When she again tried to pick up a pup Black Angel ran over and prevented her by lying on the youngster whilst, at the same time, making submissive gestures to Havoc, licking her face whilst wagging her tail insistently.

It was whilst I was watching this amusing scene that Juno suddenly began to trot briskly away, abandoning all her attempts at pulling pups from the dens. A moment later Genghis followed her, and then Swift and then, one after the other, the remaining adult dogs. I was surprised, for never before had I seen the mother initiate any movement of the pack away from the pups. Havoc was the last to leave the pups, looking after the departing pack as she stood, with some pups, at the den she had chosen. But when Juno, after leading the pack for fifty yards, returned to the pups, Lotus trotted after the other adults. Juno might have been left to do as she wished with the pups but for Black Angel. She had looked back, time and time again, as she trotted away, and when the dogs had travelled a hundred yards she suddenly turned and raced back. The other adults stopped. First Havoc, and then Swift, and ultimately the whole pack trailed back to the dens.

As Black Angel arrived, with Havoc close behind, Juno's ears were pressed back close against her head. She crouched low to the ground, her tail wagging, her lips drawn back into a scared grin. All these gestures showed extreme submission, yet the two dominant females attacked her simultaneously, biting at her neck time and time again. Juno rolled on to her side and remained motionless as Havoc and Black Angel picked up a pup together and carried it to one of the dens. However, as they were placing it inside, Juno got up and trotted towards the original den, forty yards away. The remaining seven pups all followed her. Now Black Angel and Havoc had no time for attacking the mother: for the next five minutes they frantically attempted to collect the pups and take them back to the new den. But every time they deposited one, it immediately darted away and ran after its mother. Eventually both dominant females concentrated their efforts on one pup, Sprite, picking her up and putting her in their den time and time again. But when Swift and Baskerville joined them, rushing from Black Angel to Havoc and greeting them, the added confusion gave Sprite the chance to escape and join her mother and siblings in the den of her birth.

It is worth commenting upon this episode for, whilst it seemed most confusing at the time, subsequent reflection suggests that Juno may have shown some particularly fascinating behaviour. There was nothing to indicate that she herself had been anxious to move her pups to a new den: once the new burrows were reached the most that I saw Juno do was to try and pull a pup away. As she was attacked for each attempt, I never found out where she was trying to take the pups. But perhaps she had, all the time, wanted to move her youngsters back to the old den. If so, is it possible that she initiated the move to set off hunting *deliberately*? Did she realise that, if the pack left, the two dominant females would leave too, and she would then be able to take her pups wherever she liked?

This is not quite as far-fetched a possibility as it may sound. We have seen very similar behaviour in chimpanzees. When a young male was unable to get his share of bananas, due to the presence of a number of older males (who might have attacked him had he started taking bananas), he often got up and walked deliberately away. If the larger males had had enough to eat they would frequently follow his lead, together with any other chimpanzees that happened to be in the group. Some ten minutes later the young male would return by himself and, with the field clear, sit calmly eating the bananas we gave him. This happened far too often to be labelled coincidence. And so it is not quite ridiculous to postulate that a wild dog might have employed a similar technique: it remains to prove or disprove such an idea by further observations.

The morning after the attempt to take the pups to the new dens I found all the pups playing around the den that Lotus had marked the previous evening. So it seems that, after all, the dominant females had been successful. But the pups did not remain there for long: the first move was the start of a whole succession of moves, none of them longer than a few hundred yards. And none of these new dens were inhabited for more than a few days before the pups moved yet again. On looking back over those last twenty days, from the first den move until the pack left the area, it seems that the adult dogs were getting more and more restless; they were encouraging the pups to 'find their legs', preparatory to taking up the roaming life of the adult.

There are, of course, other reasons for moving dens. One scientist, for instance, saw a mother carrying her pups, one after the other, to a new den a thousand yards away after lions had spent some while investigating the original den site. Wild dogs are usually uneasy when lions are around. Once, when I was watching a pack near Lake Legaja, the old leader stood for some minutes uttering gruff threat barks as he and the pack faced towards two lions fully half a mile away and on the other side of the lake. On another occasion I was following two female wild dogs, on their own, who seemed extremely worried by the proximity of a pride of lions. Every so often the dogs jumped up on their hind legs, completely upright, to look at the lions over the tall grass.

When Juno's pups were approximately two months old, the pack finally moved from the area. I had been expecting this to happen. For one thing the plains surrounding the den were becoming more barren and deserted every day; for another I knew of other pups which had moved off at a similar age. Yet even though I had been expecting it, my disappointment was acute when I arrived one morning at the den where I had left the pups the evening before to find that the dogs had gone. As the sun rose, nothing but the vast empty plains confronted me. There were no signs to indicate which way the dogs had gone. I had watched the pack, daily, for so long that I felt strangely desolate as I bumped my way across the plains, hoping against hope that I might, by some lucky chance, meet up with the dogs.

Soon my eyes were sore and tired from staring over the plains which, as the hours passed, became more and more hazy with the heat of the sun. Jeff searched too, but he had no better luck. I longed for a light aircraft, for only by flying for vast distances across the Serengeti, early in the morning and in the evening (the wild dogs' normal hunting times), would it be possible to locate a nomadic pack with reasonable frequency. And even then such a study would pose innumerable difficulties, for wild dog packs roam over vast areas of country provided they are not tied to one place by their young pups.

So far I have only concentrated on observing wild dogs as they move through a fairly small area, about five hundred square miles, near Lake Legaja. I know that the range of one pack includes Naabi Hill and thirty-mile distant Seronera, but this probably represents but a small part of their hunting grounds: one pack in South Africa was reported to have a range of at least one thousand five hundred square miles.

It was almost exactly a month after the Genghis pack had moved from the den area that I came across it again, as I was driving across the plains to camp one evening. I recognised it immediately, because of Black Angel and Yellow Peril with their half tails, and as the dogs came closer I identified the other members of the pack. Black Angel, with all eight pups, was bringing up the rear.

I began to follow them. We had only travelled about a mile when suddenly a hyena appeared ahead, and immediately six of the adult wild dogs rushed after it, Black Angel in the lead. When they caught up with it the dogs started to nip at its buttocks, but then instead of stopping, after a few bites, as I had seen them do on other occasions, they began a savage attack. As I drove closer I could see that they were chewing at their wretched victim's rump, and soon the hyena's growls turned into screams as it tried to escape. Every so often it stopped to bite back towards its tormentors,

1. Wildebeests, contrary to popular belief, do not always panic at the sight of wild dogs.

2. Fourteen of a litter of sixteen pups.

3. After resting, the adults and pups greeted each other.

4. Two of the adult dogs greeting a pup.

5. Tail pulling was a common game.

6. Pups showed submission by lying on their sides or backs.

7. Havoc and Swift carrying a pup.

8. Black Angel lying amongst the pups.

9, 10. The pups spent much time in play.

11. Sometimes the dogs attacked hyenas, nipping their bottoms.

12, 13, 14. Until recently it was believed that zebras were too large to be tackled by wild dogs.

16. The front dog usually grabs the tail or hindleg of the prey.

17. An adult regurgitates meat to the pups.

15. After an evening kill on the Serengeti.

18. As the pups grew older they spent more time playing or resting above ground.

19. Adult dogs often bite at each other's necks during play.

20, 21, 22. As Havoc approaches, Lotus presents her neck in submission. When Havoc puts her nose to the other's neck, Lotus shows complete submission, by lying on her side.

23. When a female comes into heat she often marks blades of grass with a few drops of urine. Here a male suitor, by balancing on his front paws, is able to mark the same blades of grass simultaneously.

24. When the pups were two and a half months old they followed the adults and started their nomadic life, passing Lake Legaja as they made their way to the plains beyond.

its mouth wide open in a horrifying grimace. But even as it bit at one dog, so another would run in from behind, and once again the hyena fled. Soon I could see blood streaming down its thighs. At last it reached a hole in the ground and, swivelling around, it lay in this depression with only its head visible, threatening the dogs with its powerful teeth. Then its tormentors moved away and returned to the rest of the pack.

I was particularly interested in the fact that Black Angel had led the attack, for on previous occasions, near the den area, we had noticed that she had a marked antipathy towards hyenas. Several times, even when she had been trotting off with the pack on a hunt, Black Angel had raced back to attack a hyena wandering near the den, biting it time and time again until it had taken to its heels and fled. Twice, when she had finished her hyena-baiting, Black Angel had been unable to locate the rest of her pack and had thus been forced to remain at the den with Juno and miss the hunt. I have often wondered whether Black Angel's hatred of hyenas had some special significance; maybe, during her youth, a hyena had bitten off the lost half of her tail.

As I drove over the plains after the Genghis pack with the three-month-old pups I wondered how far they had travelled since they had left the dens. Presently the sun set, and as it did so the almost full moon rose over the eastern horizon. Soon the dogs were no more than a row of ghostly shapes in the colourless grey country-side. Once more I felt a member of the pack, especially because, so as not to disturb the prey animals we passed, I drove without lights. I peered ahead at the moonlit plain and hoped I would notice any potholes before I drove into them.

For most of the time the dogs trotted in single file: I noticed that Havoc and Swift kept very close together, and twice they both stopped and marked the same tuft of grass with urine. After a while I realised that we had been joined by three hyenas which were bounding along, occasionally making playful biting movements towards each other. One of them (and they were all old females) grabbed the tail of her companion, and they rolled on to the ground playing together. Then they were up, bounding along again. It seemed that they were in high spirits as though anticipating some special treat.

In sharp contrast with their attack on the one lone hyena, the wild dogs took little notice of these three: even Black Angel did no more than threaten them a few times when they got too close to the pups. Hyenas always seem to be more aggressive and bold at night, and perhaps the wild dogs respect them more during the hours of darkness: also, of course, they were three as compared to one.

After we had progressed at a steady pace for about five miles the dogs stopped and lay down, arranging themselves in small groups. I drove closer, switched off the engine, and prepared to wait. For a while everything was peaceful. I was just going to pour myself a cup of coffee when suddenly, in the shadowy moonlight, I made out the fat shapes of the three hyenas, their sides touching, their noses stretched towards a sleeping dog. I watched, fascinated, as they slowly crept towards the dog's rump. I could not see what happened, for their backs were towards me and their combined bulk hid the dog from view. But all at once the dog, Yellow Peril, was up on his feet and the night was filled with loud growls as six other dogs rushed to the scene and surrounded the hyenas, each dog biting towards one of the intruders. The hyenas scattered: a minute later there was silence, and the dogs curled up once more.

Soon afterwards, to my surprise, the hyenas were back again, and once more they approached Yellow Peril, inching forwards on their tummies. This time I could see clearly and, through binoculars, I watched the noses of the hyenas gradually come to within an inch of the dog's rump. Then, in unison, each hyena slowly protruded her tongue and gave a quick lick under Yellow Peril's tail. Again the night was filled with growls as the dogs jumped up to attack the hyenas, driving them away once more. Afterwards the dogs lay together, adults and pups, all in one big heap. Yellow Peril paused before joining the others, squatting to defecate. Hardly had he moved from the place to resume his interrupted rest before one of the hyenas had rushed to the spot and eagerly consumed the dropping.

Previously I had seen hyenas eating the droppings of wild dogs; indeed, it seems that they are a special delicacy, but I had not realised that a hyena would actually lick the bottom of a wild dog to satisfy this strange craving.

About an hour later Genghis suddenly got up and almost immediately the rest of the pack was on its feet and the dogs were rushing around, squeaking and twittering, wagging and licking, in the frenzy of the greeting ceremony. They then set off again, and soon I was once more driving across the ghostly plains. The three female hyenas were still with us.

Soon the dogs paused, staring into the darkness ahead, and focusing my binoculars I made out the shapes of a small herd of Thompson's gazelles. Then the pack set off slowly towards their prey. As the chase started I pressed on my accelerator, but I had only driven a couple of yards when I crashed into a hole. By the time I had reversed out I was alone in the night. Quickly I began to zigzag over the plains and, after a while, I met one of the hyenas. She, too, was looking rather lost, but I knew her sharp ears would pick up the slightest sound, and so I followed her. It was lucky, for only a minute later she started to run, and within seconds the other two hyenas appeared and ran along beside her.

Presently I saw a dark mass ahead, silhouetted against the paleness of the dry grass, and I knew the dogs had been successful. Quickly I braked, and grabbed my binoculars. The hyenas did not slow their pace, but rushed straight in amongst the dogs and, a moment later, all three were lying on the carcass, covering it with their large bellies. And there they lay, rump to rump, twisting and turning and giving their hysterical giggling sounds as they bit towards the dogs who, in turn, rushed forward to nip the hyenas. But despite the nips the hyenas held their ground and, after a few minutes, when other hyenas began appearing from the darkness, the dogs abandoned their kill and trotted off into the night.

It was not long before the dogs chased another Thompson's gazelle. This time, as no pothole lay in my path, I was able to keep up with Juno, Yellow Peril and the pups. This little group ran way behind the rest of the pack. After a while Juno and Yellow Peril stopped and seemed to listen intently. Then they ran off for a short distance, only to stop and listen again. It was obvious that they had temporarily lost track of the others. And then, from the darkness, came the strange high-pitched flute-like 'hoo' of the lost wild dog: a few minutes later three other wild dogs joined us – Genghis and two other males. They too seemed lost. Suddenly, however, their large ears must have picked up some sound inaudible to me, for they started to run fast, and I had difficulty in keeping up with them.

When we arrived with the remainder of the pack, Genghis, Yellow Peril, Juno and the two males rushed to tear at the kill, but the pups stopped and lay down some thirty yards away. Normally wild dog youngsters rush up to feed as soon as a hunt is successfully terminated; whilst they take their fill the adults actually stand back watching. But when I drove closer to the kill I found that there were two hyenas there, trying, unsuccessfully, to share the meal. This, perhaps, explained the pups' reluctance to approach for, although I have never seen a hyena catch one, I have often seen wild dog pups avoiding these strong-jawed predators.

When the adults had finished feeding several of them regurgitated meat for the youngsters. And then the pack trotted off once more across the plain. Soon, however, some flashes of lightning lit up the sky and, not long afterwards, the moon disappeared behind the clouds that heralded the coming of the short rains. Then I could follow the dogs no longer and, regretfully, I switched off the engine, curled up in my blanket in the back, and awaited morning. I knew I should not find the pack the next day.

Over the years, by means of scattered observations on a number of packs, I have been able to learn something of the gradual development of wild dog pups after they have left the den. One such pack was of great interest since it comprised eight pups, of about three months old, and only three adults, two of which were males. During the one night that I was able to observe the pack the pups were left on their own during the hunt. When the adults raced off after a Thompson's gazelle the pups followed for about one hundred yards and then lay together in a heap. After five minutes they suddenly stood up and stared in the direction from which we had come: they seemed worried and soon I made out the vague shape of a hyena approaching, sniffing along the ground. The pups ran off a short way and then stopped, looking back towards the place where they had seen the hyena. Soon it appeared again, moving slowly, sniffing along the trail of the pups. The grass was tall; it seemed that the hyena did not see the pups and they ran off again. I was interested to note that, although the hyena approached four times, and four times the pups ran off, they never went farther than thirty yards or so at a time. Had they gone farther the three adults, who reappeared fifteen minutes after the start of the hunt, might have had some difficulty in finding them. As it was, the pups saw them immediately and soon the pack was running fast over the moonlit plain, the adults leading, until we came to the carcass of a Thompson's gazelle. The adults had obviously eaten well before returning for the pups, for a good deal of meat had gone: now they stood aside as the youngsters fed. If hyenas had taken the kill before the pups' arrival on the scene, the adults would, undoubtedly, have regurgitated to the youngsters. But, with eight pups to feed, all would have gone hungry and the adults might well have made a second kill. As it was, all the dogs seemed contented when they finally trotted off; they spent the remainder of the night sleeping in a close group.

In sharp contrast to that pack, I watched another comprising twelve adults and but one pup, of approximately five months old. These dogs killed two adult Thompson's gazelles in one night. During the first hunt the pup ran along behind with three other adults but, before the kill had been made, these adults also left it behind. For about five minutes the pup seemed to be lost: it kept standing and staring in all directions. But then two adults appeared from the darkness, the pup raced over to them and greeted them frantically, and the three ran on to the kill. There, having greeted several other adults, the pup pushed into the carcass and, within a few moments, all except for one adult had moved away from the kill. That one was the pup's mother. Twice, when adults approached the food, their tails wagging, the pup made quick darts towards them, and they moved away again. During the second hunt, about four hours later, the pup remained close to a group of adults in the rear and when it came up to the kill it made no attempt to push in and feed. After this it seemed that the hunger of all thirteen dogs was assuaged, and the pack slept peacefully for the rest of the night, stretched out under the moon.

One other pack must be mentioned. This one comprised five adults, of which only one was a male. This was somewhat unusual since, in most packs, males outnumber females. In addition, there were eleven youngsters of under a year old. When the pack hunted the first evening these youngsters joined in the chase but stayed some way behind the four females, together with the adult male. When the kill had been made the youngsters ran up and took part in the disembowelling of the prey, a calf wildebeest, but then, rather to my surprise, the adults moved away and watched them feeding. I had not imagined that the grown dogs of a pack would stand aside when their youngsters were so old: indeed, save that the adults of this pack were of slightly thicker build, there was little difference in the size of the two age groups.

When the wildebeest calf had been finished the adults immediately started to hunt again. There was no shortage of prey, for the pack was in the midst of the wildebeest migration, and nearly all the cow wildebeests were accompanied by young calves. The first hunt had been very quick, but the second was longer. Several times the pack of wild dogs ran towards different herds and then stood watching as the wildebeests ran past. Soon the prey were panicking, and Jane, who was in the car with me, suddenly noticed that several large herds were galloping straight through our camp. From where we watched the dogs we could just see the green canvas of two of our tents and we were almost tempted to leave the wild dogs and race back to see that all was well. However, Jane's mother was there, and we felt convinced that the thundering hoofs and the huge dust clouds, golden in the sunset, would quickly warn her to take Grublin to the safety of the Volkswagen. Which, of course, was so.

The following day we saw the same pack hunting and again the prey was a calf wildebeest. This time a group of orphaned calves cantered almost up to the pack and the young dogs made two kills themselves, helped, in one instance, by one of the adult females of the pack. She then stood aside, with the other four adult dogs, and watched as the youngsters fed. When nothing was left save a few bones the five adults made a kill of their own – the youngsters made no attempt to join in.

So far I have not been able to watch the transition between the youngster, with its prerogative of feeding first, and the adult. Presumably this would only be possible if one could keep a pack under constant, or almost constant, observation, over several weeks. What a task! For, as I have said already, a pack of wild dogs may roam over one thousand five hundred square miles or even farther. Moreover, it is assumed by many people that, within this vast area, a pack has no regular routes, no preferred areas, but that it will simply wander where prey is the most plentiful, varying its roaming from month to month, from year to year.

The Genghis pack, however, gave me two clues which indicate that this may not be quite true – and certainly proved to me that wild dogs may know at least some areas in their range very well indeed. The first hint was provided when I was following the pack at night, during the short rains. After making a kill and feeding the pack moved on. Genghis led the dogs away across the plains for about four miles, up a slight rise and down to a pool of water. If, after drinking, the dogs had continued in the same direction I should have presumed that their coming across this pool was a mere coincidence, but instead, when all had drunk, Genghis led the pack back over the four miles they had travelled to get there. Had they, perhaps, smelt the water? I wondered at the time, but six months later, in the middle of the dry season, the same pack gave me evidence which suggests that they actually know the whereabouts of waterholes in some parts of their range.

As I followed the pack across the dry countryside, I saw that Genghis made a slight change of direction and led the others to a dry waterhole. The dogs paused for a moment, sniffing around, and then trotted on. I thought nothing of it. But six miles farther the dogs suddenly increased their speed slightly as they climbed a small ridge. They stood looking down at another dry waterhole: I got the impression that they were disappointed and, for the next ten minutes, they moved around sniffing at the dry sand. I feel convinced that the pack had gone there specifically to drink: the waterhole had not been visible until we were a few yards from it, and as it contained no water, neither sight nor smell can have contributed to the sudden quickening of pace I had noticed as the dogs approached the place. It is interesting that whilst both wild dogs and hyenas will drink daily and spend hours lying in water or mud if it is available, they seem to be able to go without water for long periods of time.

It is possible that, within the vast hunting grounds of a wild dog pack, a certain area is preferred for the raising of pups. Certainly different females of the Genghis pack, for three years running, raised their young in the vicinity of Naabi Hill. True, two of the home dens were at least ten miles apart, but when the entire range is so huge this is a relatively short distance. I know of another pack which has twice raised its pups in exactly the same spot, and Dr Leakey tells me that a pack has returned to Olduvai time and again to rear its young. One might call such a favoured place the 'breeding ground' of a pack.

Obviously a pack of wild dogs cannot expect to keep the whole of its extensive hunting grounds to itself: the ranges of a number of packs must and do overlap. On two different occasions, with two years between, I saw the Genghis pack chase smaller packs away from its breeding grounds. Another small pack which hung around in the breeding grounds of the Genghis pack for some ten weeks (a most unusual length of time for a non-breeding pack to remain in one fairly small area) left the district completely the day the Genghis pack returned for a while.

At other times friendly relations exist between different packs. Three times other dogs joined the Genghis pack, moving and hunting with it for varying lengths of time. I saw none of the actual meetings, but when I came across the enlarged pack, relationships between the old members and the new temporary ones seemed relaxed.

Why are some wild dogs chased away and others permitted to join in the life of the pack? I suspect that this may have a good deal to do with the previous history of the dogs concerned. One pack, for instance, in the Mikume National Park, numbered some forty a few years ago. At that stage it started to split up into three separate groups which, from time to time, still met up, moved around and hunted together. Presumably recognition and tolerance of previous pack members would persist over long periods of time: after all, if the wild dogs can remember individual waterholes in their hunting grounds, it should not be difficult for them to remember old hunting companions. Ultimately, though, the original dogs, who knew each other before the pack split, will die and the younger ones will gradually become less and less familiar until, eventually, friendly relations may give place to aggression when two such groups meet up. But this, of course, is mere speculation, and many years of research lie ahead before factual answers can be substituted for guesswork.

When a pack of wild dogs enters an area occupied, at that time, by another pack, it will normally receive ample warning of the presence of the others through its sense of smell. In addition to detecting, from urine or droppings, the fact that 'other dogs' are there, the intruders will probably be able to decipher the scent signals more accurately: they may read that 'those particular dogs' are there. Thus, if necessary, the intruding pack can discreetly withdraw without ever meeting the others.

Scent-marking, which occurs in many mammals, is often associated primarily with the marking out of the boundaries of a territory, a territory being that part of an animal's home range which it will defend from strangers of the same species. Since wild dogs, so far as we know, have no territory as such, they cannot patrol its boundaries, posting warning signals, as many species of carnivore do. But occasionally, when a pack is hunting, one of the dominant members will pause by some tall tuft of grass, raise a hind leg and deposit a small trickle of urine there. Havoc and Swift, in the Genghis pack, were most liable to mark in this way, and very occasionally one of the other dogs would add a contribution of its own. Yellow Peril always amused me when, on rare occasions, he wandered over to such a spot, for his behaviour suggested that, to him, it was a most unimportant affair. After sniffing briefly at the tuft he would lift his foot no more than an inch from the ground and normally deposit no urine at all. At the most, as if by mistake, he contributed one or two drops to the message.

It has been suggested that scent-marking has evolved from the uncontrolled production of urine and droppings by animals that were frightened in a new environment or a new situation. Subsequently, so it is argued, the new situation became less intimidating because the animal was surrounded by its own familiar smell. Since then, of course, scent-marking has gradually developed a whole variety of other functions: undoubtedly it helps dispersed members of a group to find each other; it can indicate ownership of territory, the presence of a female in heat and whether or not she has found a suitable mate.

In wild dogs, as in domestic dogs, it appears that there is an increase in the marking behaviour of a female just before and during her period of sexual attraction. This was most obvious in a pack of four wild dogs, two males and two females, that we observed quite regularly for nearly ten weeks. The behaviour was not only interesting, but most amusing to watch.

The dominant female, advertising her condition wherever she went, was closely followed by the dominant male. Each time she marked he rushed over and marked in the same spot. He was presumably attaching a little warning to her advertisement, announcing to any other prospective suitor who might pass by that this female was being attended to and that any interference would cause big trouble. And not only did he attach his warning, quite literally, to the *exact* spot where she marked, but frequently was in such a hurry to do so that he marked at precisely the *same moment* as she did. This, of course, was no easy matter: imagine two dogs, standing side by side and almost touching, as the male tries to hit the identical blades of grass which his lady is scenting. If he raises a hind leg sideways at the same time as she does they will push each other away. And so this male performed an amazing acrobatic feat, fully worthy of a circus dog: he stood on his front feet, his hind legs up in the air, his body vertical. His sense of balance was superb – he could even walk a few steps forward. By doing this he was able to spray the same blade of grass at the same moment as his female: their perfumes were inexorably mixed and none who passed later could doubt but that the lady fully accepted the proximity of her suitor.

We have seen two other females in heat, in two other packs. In one of these packs we did not know the dominance order of the different individuals: during the few days that we kept up with the dogs only one male mounted the female. In the other pack two males courted and possessed the female in turn: neither was the dominant male. This female marked frequently, and when she had moved away from the spot the dominant male moved over to mark there too. But the other six males, including her suitors, merely sniffed at or rolled on the spot.

Each of the three bitches that we have watched during her period of heat was closely followed by her suitor, who often walked along with his nose actually touching her rump. Each of the suitors frequently headed his bitch away if she approached other males too closely, pushing against her with his muzzle or the side of his body. When the female stood, the male often rested his chin over her shoulders or back, and when she lay he lay close beside her, usually leaning possessively against her body.

Courtship was somewhat brusque: the male stood up, after he and his female had been resting, and proceeded to paw and scratch at her with one paw. Eventually, when she got up, the male mounted her. One fact should be mentioned here: it has been stated that the wild dog differs from the other members of the Canidae (or dog family) in that there is no tie after mating. In fact, on all occasions when we saw successful mating, there was a tie. Unlike domestic dogs, however, which may be tied for up to twenty minutes, the tie in wild dogs was usually no more than fifty seconds, although on a few occasions it lasted three and once five minutes.

One thing which interested us, right from the start of our study, was the dominance hierarchy, or pecking order, in a pack of wild dogs. It has been stated often that no such thing exists, yet it will already have become obvious to readers that it does. In all the packs which I was able to watch for more than a week I saw some indications of a rank order at least amongst some of the individuals. Why, then, have other observers not come to the same conclusion? The situation may perhaps be compared to a family with sensible parents who understand one another, and well brought-up teenage children who get along with their elders. For days on end the parents may have no occasion to reprimand a child with more than a gentle word. The mother or father may order a child to do something, thereby enforcing their dominant roles, but if an observer did not know the language he might not realise what was going on. Only when a situation arose which resulted in a disruption of everyday behaviour, such as a quarrel, would such an observer have a chance to work out the status of the different individuals of the family.

And so it is with dogs. The members of a pack usually know each other well, and it is rarely that a situation arises which induces one or more of the dogs to assert their dominance. And only after watching many such situations was it possible for me to interpret the dogs' 'language' of ear, tail and body positions.

It was much easier, in all the packs but one, to detect dominance interactions between the females, and it was the females of the Genghis pack who taught me what to look for. Havoc, as we have seen, was a high-ranking female. When she was being slightly aggressive to one of the other females she merely stood, her head erect, her ears pricked, her tail hanging down and quite immobile. If she felt more aggressive she would lower her head slightly and approach the other with her head parallel to the ground, her ears lowered sideways, and her tail, again, immobile. She might point her nose towards the neck of the subordinate or, if the situation warranted it, actually bite the other in the neck.

The low-ranking Juno was alert to the slightest signs of aggression in any of the other dogs. She would respond by lowering her head, pointing her nose to the ground and pressing her ears back close to her head. As the dominant dog approached she wagged her tail more and more frantically, and often crouched low to the ground. When she was actually threatened she flopped right down on to the ground, usually rolling on to her side or back and spreading her hind legs. Sometimes, too, she pulled back the corners of her lips, exposing her teeth in a grin of fear.

Black Angel, lower-ranking than Havoc, was a more dominant dog than Juno, and less submissive, usually, during her interactions with Havoc. Instead of cringing and grinning in fear, she usually licked and nibbled the lips of the dominant female, or rubbed her chin over the other's head time and time again. But she too lowered her ears and wagged her tail vigorously from side to side.

One most interesting gesture of submission in wild dogs, and one which is shown equally by very low-ranking and higher-ranking individuals to their superiors, is the presenting of the neck, when the subordinate slightly turns his muzzle away and presents his curved neck to the dominant animal. By so doing he is, of course, turning away his only weapons, his teeth, and the posture, therefore, demonstrates clearly that his intentions are not aggressive. Frequently, though, this presentation of the neck seems to lead to a conflict of behaviours in the submissive animal: on the one hand, the desire to appease, by turning the teeth away; on the other hand, either a desire to show friendly behaviour, by licking the other's lips, or a desire to be able, if necessary, to defend itself – both of which involve turning the muzzle *towards* the superior. These conflicts appear as a series of head shakes as the 'towards' and 'away from' movements are rapidly suppressed and alternated.

Dr Lorenz, in his extremely well-known book *Man Meets Dog*, describes neck-presenting as a submissive gesture in the wolf and the domestic dog. Since Lorenz wrote this he has been attacked on the accuracy of his observations. Dr Schenkel, who has watched American wolves, silverback jackals and domestic dogs, claims that it is only the dominant individual which presents its neck. For one thing, this puts it in a good position in which to swivel its hindquarters round to slam them into its opponent (in the same way as practised by the golden jackals) and, for another, its confident stance inhibits the subordinate from biting. Dr Shenkel suggested that Dr Lorenz might have confused the dominant with the subordinant animal in such interactions

But when I saw this gesture in the wild dogs, so unmistakably the sign of submission in low-ranking individuals, and when I saw the same gesture, quite unmistakably, in subordinate golden jackals, I began to think again about Dr Lorenz's observations. And then I met Dr de la Fuente who has had experience with *European* wolves. He looked surprised when I asked him if he had seen a neck-present in a submissive wolf and replied that it was amongst the most common submissive patterns.

Perhaps, after all, therefore Dr Lorenz was correct: perhaps the wolf in Europe differs more from the wolf in America than Dr Schenkel was aware. All I know is that neck-presenting, in a submissive wild dog, is one of the most common gestures of subordination that I have seen in any of the packs I have watched so far.

One fact which, more than any other, has probably contributed to the generally accepted idea that there is no rank order in a wild dog pack, is their greeting behaviour. For during this brief and chaotic ceremony it is usually impossible to sort out any social status in any of the dogs unless you concentrate on watching one individual only. Even then, the rank of that individual is often not revealed for it seems that this behaviour is so ritualised that each dog usually shows the same patterns as the companion he is greeting. So that when Havoc ran along with tail and ears erect, then Juno did the same. Both might lick each other's lips and, as they did so, lower their heads towards the ground, farther and farther, as though competing as to which could go the lowest. It may well be, as I have hinted, that this ceremony demonstrates a ritual subordination of the individual to the interests of his pack, and thus serves as a mechanism to ensure the successful co-operation of each dog in the hunting unit. I was interested to learn that a wolf pack, which has a very clear-cut dominance hierarchy, demonstrates a pre-hunt greeting ceremony that is strikingly similar to that of the wild dogs.

Another factor which has contributed to the 'no dominance' fallacy is that, when the dogs are feeding at their kills, it is unusual for squabbles to break out. But here again, provided one is able to watch a pack over a long enough period, it will normally be possible to find some indication regarding the rank order within the group, even at a kill.

In one pack, comprising nine adult dogs, two individuals were invariably chased away by more dominant animals whenever they approached to feed. One of these unlucky dogs was a young male, so lame that he could only run on three limbs, and I was surprised to see him driven off, knowing that a pack will sometimes regurgitate food to a crippled member.

I was only able to keep up with this pack for a few days; during that time the dogs killed three Thompson's gazelles and persistently kept this lame individual and one old male away from the meat. The lame dog fared the worst for, so far as I could tell, he only managed to get one fairly small bone from one of the kills. And yet, despite this seemingly intolerant behaviour of the pack towards him, there was no indication that he, or the old male, were outcasts in any other respect. Once, after chasing a hyena, some jackals and countless vultures from a wildebeest carcass, the entire pack waited for over an hour whilst the lame male struggled to pull a few dried-out shreds from the bones and, when a hyena approached the carcass too closely, several dogs rushed over and helped the lame male to chase it away. And on another occasion, when he was threatened by a small group of belligerent bull wildebeests, the pack again rushed to the lame male's assistance and chased them off.

Looking back over the episode I suspect that the somewhat unusual behaviour may have been due to the fact that one of the two females was in heat. It was her suitor who most frequently drove the two dogs away from the kill, and it may well have been a response to their approach to his female, rather than to his food, which prompted his aggressive behaviour.

The only other examples of persistent aggression amongst adults at a kill occurred in the pack of four dogs which Jack, Roger and I were able to observe regularly for nearly ten weeks. During the first few weeks of this period the lower-ranking of the two females was constantly chased away from every kill, and often attacked by the other female. Added to which, one or both of the males often joined the top-ranking female in her attacks, rather as pups will gang up to mob one of their siblings when it gets the worst of a rough-and-tumble game. When the dogs were not feeding, however, the dominant male quite frequently hurried to defend the subordinate female if she was threatened by the other one, either putting his body between the two and gently pushing the dominant female away, or actually threatening her with his nose pointed stiffly into her neck. After the first few weeks of our observations, however, the dominant female became, for some reason, less aggressive and allowed her erstwhile enemy to feed in peace.

We learnt a great deal about the patterns of dominance from this small pack, mainly because, after we had watched them for over a month, the dominant, or number one female, came into heat. When this happened the dominant male, as I have mentioned earlier, seldom left her side. Even before this the number two male had shown great respect for his superior: now he kept well out of the way of the courting pair. Occasionally, though, it seemed that the number one female wanted the best of everything – from everybody – and every so often she would wander off in the direction of the subordinate male. Her suitor was constantly alert for this fickle behaviour, and even when the pair was sleeping he always placed himself so that he was between his female and his rival. If she moved, number one male was up in an instant, hurrying to position himself between his lady and the other male. If she got too close to the other he would gently push her away with the side of his body. Sometimes she ignored this signal and pushed past him (always in the most casual way), but only if she got to within a few yards of the subordinate male did her suitor turn his attention to the prompt removal of his rival. He did not have to exert himself to achieve this, for his wretched subordinate became uneasy as soon as he noticed the seductive approach of the female, and often, indeed, moved away from the temptation of his own accord. If he had not already moved, then the slightest threat from male number one was sufficient to send him hurrying to a more discreet distance.

At the height of the dominant female's period of sexual attraction the dominant male had little chance to rest. His lady only needed to shift position a few feet for him to become instantly concerned. Jumping up, he rushed over to her, licked her, rushed to where she had lain and marked the spot carefully, and then hurried over towards the subordinate male. Always the latter was lying at least fifteen yards away during this period, but at the approach of his superior he moved even farther away: the frenzied suitor merely marked vigorously the place where his rival had been lying, before rushing back to lick his lady once more and then lie, in uneasy rest, at her side.

The relationship between the two top-ranking dogs and the subordinate female reminded me very much of that which existed between Havoc, Black Angel and Lotus in the Genghis pack. In that triangular set-up, Lotus often approached Havoc as though she felt that proximity with the dominant female would strengthen her own status. At which Black Angel, who obviously felt the same, hurriedly went to place herself between the other two. Thus in the pack of four dogs, the submissive female constantly moved so as to be close to the number one male, and the dominant female instantly responded by pushing between the two. This was particularly apparent when the dogs were resting. The number one male and number one female nearly always lay together: the number two female usually lay with them. But she never lay with the dominant female unless the dominant male was there – perhaps she relied on him to protect her. But neither would she lie with him unless the other female was already there beside him – possibly to have done so would have antagonised the other female and provoked an attack. Sometimes I am amazed when I realise how subtle a subordinate animal learns to be in order to avoid trouble.

I am also amazed – and here I run the risk of scientific disapproval for crediting animals with human emotions – at the extent to which an animal may, apparently, nurture a desire for revenge. Let me give an example.

It all started when I met up with the Genghis pack about two months after Juno's pups had left their home dens. Once again this pack had returned to its breeding grounds and, as I was working on cheetas, I met them quite by accident. Nevertheless, I immediately gave up my other work and determined to stay with my wild dogs as long as possible – for nearly a week, as it turned out, for there was no moon at all, and each morning I found the pack resting very close to where I had left them the night before.

The first evening the pack made a kill after an exceptionally long chase. The pups, just over quarter-grown, lagged far behind and, when they finally came up to the kill – a Thompson's gazelle – there was hardly any food left, and they began begging from the adults. Havoc responded immediately, regurgitating some meat. Black Angel also ran quickly to the pups and started to regurgitate close by. Exactly what prompted Havoc's attack I am not sure – possibly she thought Black Angel was about to eat the meat that she, Havoc, had regurgitated. In any case, she suddenly flew at Black Angel, biting her time and time again in the neck. Quickly Swift joined her in these attacks.

And then, suddenly, Lotus noticed Black Angel crouched under the two dominant dogs; immediately she hurried over and joined in. But whereas Havoc and Swift were not biting very hard, Lotus instantly drew blood: she bit into Black Angel's neck time and time again, until one of her victim's forelegs was dripping with blood.

However, if Lotus had thought to dominate Black Angel in that one minute, she was mistaken. When Havoc and Swift moved away, Black Angel promptly turned to get her revenge. For twenty minutes she chased Lotus around, and the latter, despite her moment of triumph, took good care to keep well out of Black Angel's furious way.

It seemed that this incident completely threw the normal pattern of dominance between the four females. For the next few days Black Angel was particularly aggressive towards Lotus, and disagreements between those two and Juno were quite frequent when the pack was feeding. It was particularly noticeable that Juno was far less submissive and cringing in demeanour than when I had watched her previously. Possibly, now that she no longer had to interfere with the more dominant females' enjoyment of the pups, she was less liable to sudden threat and attack, and had gradually gained confidence. Nevertheless, I was not prepared for the spectacular fight which took place between Juno and Black Angel.

The circumstances were slightly involved but should be explained in brief to account for Havoc's absence when the fight took place: had the dominant female been present the outcome might have been different. It was when a small pack of five stranger wild dogs appeared, trotting over Naabi Hill, that the Genghis pack split up: Havoc and Lotus and most of the males went racing off after the strangers, whilst Black Angel and Juno, together with Yellow Peril and one other male, remained behind with the pups.

During the previous few days I had noticed that a strange bond seemed to have sprung up between Black Angel and Yellow Peril, the two dogs with only half a tail each. It seemed an unlikely sort of partnership, yet constantly they trotted about together, or lay curled up side by side. And Yellow Peril had performed some, for him, quite energetic marking, after noticing Black Angel scenting a tuft of grass. I mention this only because of its possible connection with the cause of the fight.

It started when the pups were lying in a heap and the four adults standing, looking in the direction where the rest of the pack had vanished. Juno paused, as she moved past Yellow Peril, and he made some licking movements towards her. Perhaps Black Angel, who looked over at that moment, considered this an infringement of her special relationship with the old male: at any rate, she bounded towards Juno, her ears up and her tail arched aggressively over her back. But Juno, instead of running away, rushed to meet her, and a moment later both females had reared upright, supporting themselves with their front paws on each other's shoulders. The position was no different to that which I had watched so often during adult play, but this was far from playful for Black Angel had a fast hold on Juno's throat. Seconds later, however, Juno managed to wrench herself free and then, in her turn, to grip the other's throat. Black Angel, at the same time, seized her opponent's ear and soon tore off a jagged piece of skin, but Juno did not for an instant let go of her grip on Black Angel's throat. Suddenly, however, Black Angel jumped up off the ground with a lightning movement and, whilst in mid-air, managed to twist her body and kick towards Juno who let go and staggered backwards.

It was a momentary respite; a few seconds later the two reared up again and, for the next few minutes, first one and then the other managed to get a brief hold on the other's throat. By now small patches of blood were soaking the neck hair of both dogs, and they both had pieces torn from their ears. And then, about five minutes from the start of the fight, Juno managed to get a really good hold, her teeth sinking deep into Black Angel's throat. Soon the blood was flowing into Juno's mouth and dripping on to the ground.

Black Angel was my favourite wild dog, and I watched in horror as she made fruitless attempts to pull away. Soon her movements were noticeably feeble and, as the seconds went by, she started to sink to the ground, her head drooping to one side. Suddenly she gave a heart-piercing shriek and, as she collapsed on to the ground, Juno let go and stood panting over the motionless body, the blood still slowly dripping from her mouth. Then she trotted a short distance away and sat down to lick her wounds.

I felt convinced that Black Angel was dead but, after a moment, she cautiously raised her head. Instantly Juno rushed back, and Black Angel lay quite still again, presumably indicating that she accepted defeat. Again Juno trotted away. For the next ten minutes Juno hastened across each time Black Angel moved, but finally she allowed her to get to her feet. Then Black Angel excelled herself with submissive gestures, licking the other's lips and mouth, wagging her tail, pressing her ears against her head, and drawing back her lips in the grin of submission. She was making it quite clear that, for the time being anyway, she fully accepted the other's superior status. This, apparently, was sufficient: death had been close, but Juno had not demanded the ultimate penalty.

During the fight Yellow Peril and the other male had hovered in the background with the pups. Once Yellow Peril had approached, as though to join in in some way, but perhaps the very concentration and ferocity of the battling females had kept him away. He still trotted beside Black Angel when the small group moved off, following the direction taken by the rest of the pack. I followed as far as I could, but soon darkness set in: the next day the Genghis pack had moved on.

When I next met up with the pack, one and a half months later, Juno was still number two in the female hierarchy, second only to Havoc. Things seemed to have settled down, for there were no more squabbles amongst the females when the pack fed, nothing to indicate the stressful time through which Black Angel must have passed. By then the short rains had started and the plains were green and dotted with herds of wildebeests and zebras. The adult dogs spent much time in play. They chased each other and the pups over the plains, Swift somersaulted over Havoc, Black Angel and old Yellow Peril jumped up at each other like youngsters, the pups tugged one another's tails.

There were still eight pups, and in this respect the pack was fortunate for many youngsters vanish during the early days of their wandering lives. But one big change had come over the pack. Old Genghis, the leader, was no more.

Golden Jackals

the Daring Scavengers

Golden Jackals
the Daring Scavengers

It was getting on for evening, with the sun still hot in the sky, as I watched the pair of golden jackals trot, one after the other, across the short green grass of the Ngorongoro Crater plain. About once a minute the male, who later became known to us as Jason, stopped by a tuft of grass and marked it with his scent, raising his leg after the manner of a domestic dog. Then he trotted on. When Jewel, his vixen (as *we* call the female jackal), reached each of the tufts she added her own scent to that of her mate and then followed him.

This was a courting couple and, though I did not know it at the time, they were marking part of the boundary of their territory – the territory where, later, they would raise their cubs and where, quite possibly, they had raised cubs before. All at once I saw Jason stand rigid with his hair on his neck, back and tail erect. Jewel stopped beside him and the pair stared intently in the same direction. Then I, too, saw the third jackal lying curled up, seemingly asleep. Jason's lip curled and, snarling, he went and stood over the intruder in a threatening attitude, his back arched, his head held low with ears down and mouth wide open, showing his white teeth. The third jackal seemed to press itself low into the ground.

All at once they were fighting. It was impossible to follow the exact details so fast did the combatants move, but as the intruder made a bid to escape, so Jason darted forward to bite its neck and then, for a few seconds, both of them stood on their hind legs, biting towards each others' faces and necks. Then the intruder was off, streaking away across the short green grass, Jason, followed now by Jewel, in close pursuit. Soon, however, the routed jackal had apparently run far enough from their territory and the pair stopped their chase and stood together, watching him go.

For a moment Jewel stood briefly on her hind legs, her paws on her mate's shoulders, and then, starting around one ear, she began to groom him. After the first few minutes Jason seemed to relax, first sitting and then lying with almost closed eyes as his vixen nibbled at his grey-gold coat. For twenty-five minutes she groomed him, moving around as he lay in order to reach different parts of his body, pausing from time to time only to nibble or scratch at her own fur. Then she stopped and Jason groomed her for about five minutes before he moved away and, curling himself into a ball, appeared to sleep. Jewel glanced at him and then lay down herself.

It was the mating season for golden jackals in the Ngorongoro Crater, and for the next few days I watched Jason as he courted Jewel, approaching her with his tail held stiffly out behind him, his ruff erect, his ears pricked forward. But usually, as he moved to sniff her rump, she swung round and snapped at him before trotting away. For the most part the golden jackal hunts on his own, but during the mating season a courting couple, such as Jason and Jewel, rest and hunt together. Often they groomed each other; Jewel groomed her mate for very long periods, but Jason too often groomed her for fifteen minutes or more at a stretch. Occasionally, when they both stopped to sniff the ground together, or mark, or roll on some dead mouse or scrap of meat, one or the other would stand briefly with its forepaws on the other's shoulders – behaviour I have only seen during courtship. I only saw them mating once, very early in the morning, and I guessed they usually preferred the hours of darkness.

We were only visiting the crater for a short while then, but two months later we were back for a six-month visit, this time with two students to help with the observations. It took many hours of hard work to get our camp set up around the Munge cabin. There were overloaded cars to unpack, tents to erect, things to store away. The car trailer, after being unpacked, served as a store for tinned food, out of reach of prowling hyenas for, surprisingly, these creatures seldom jump up from the ground during their nightly scavenging around a camp. If they get the chance, however, they will carry off unopened tins of food which they chew until little remains of the original shape.

As usual Grublin delayed operations. He was only a crawling ten-month-old, but he invariably managed to creep to all the most inconvenient places. If a tent was ready for erection then Grublin was sure to be on top of it; if a neatly packed box was left for a moment on the ground we would soon find Grublin inside it and all the contents strewn over the grass.

Eventually, however, things were more or less ready and there was time to drive up Scratching Rocks Hill before sundown. This is a small hill about a mile from the cabin and serves as an invaluable viewpoint across the flat crater plains. Slowly the sun sank as the herds of wildebeests passed the hill in black lines on their way to their nightly feeding grounds. Some young hyenas played near their den; a flock of white cattle egrets settled in a large bare tree for the coming night; and over in the green reeds of the marsh a small herd of elephants gradually moved out on to the plains. As we sat there the sky turned golden and then red and finally darkened.

For my study I wanted information on the development of jackal cubs and the changing relationships within the family as they grew older. It seemed almost too good to be true that Jason and Jewel, whose courtship I had watched, should have a litter of exactly the right age. When first I spotted them, standing close to their den, the four cubs were still dark, almost black, and when one walked I saw that it still wobbled on its short legs. After a while I drove a little closer; one cub took a lurching step away, bumped into its siblings and, like ninepins, one after the other they lost their balance and toppled down into the den. Soon, though, they came out again. I could see that their eyes were still misty blue and I guessed that they had not been long open. We know, from zoo records, that the eyes of a jackal cub open when it is about ten days old. These four cubs must have been some two weeks old.

Later in the morning their mother, Jewel, arrived at the den. She was wary of the Land-Rover at first, so I backed away. Then, after staring at the car for a few minutes, she trotted to the entrance of the burrow, put her head down and gave a few soft whines. The cubs immediately appeared above ground, followed Jewel to a spot some ten feet from the den entrance, and suckled. In order to reach her nipples they stood on their hind legs and supported themselves with their front feet against her belly. Sometimes the paws of one cub slipped away and it hung on to her teat for dear life whilst frantically trying to regain its lost support.

I had expected that Jason and Jewel would take it in turns to stay near the den whilst their youngsters were so small, but although both parents spent a number of hours each day curled up close by, there were long periods when the cubs were quite on their own. For the first few days that we watched them the four spent a good deal of time down the den. When they did come up they wandered about, sniffing the ground, or played gently, crawling over and biting at each other. On the third day I saw two of them try to jump on each other during one of their games. They were inexpert, often misjudging distances and even direction. When they landed they usually tumbled over – sometimes down into the den. And when they did manage to land squarely on all four feet, they stood for a few moments as though waiting for the wobble to leave their legs before taking another step.

It was early one morning, five days after I had made the acquaintance of Jason's cubs, that I first saw a hyena investigate the den. At the time the cubs were down in their burrow and the parents were curled up and sleeping separately, each some twenty yards away. I watched as the hyena slowly headed towards the den, pausing from time to time and sniffing the air. When it reached the entrance it pushed its head right down. I imagined the four cubs below seeing the last gleams of daylight disappear and undoubtedly smelling the hyena's rank breath. I wondered whether the hyena would start digging and unearth the cubs.

All at once there was a loud growl. The hyena leapt backwards, twisting round and snapping its jaws at Jewel, who had rushed over to bite the hostile rump protruding from her den. Her movements were like quicksilver – darting away, she turned to face the hyena, raised her nose to the sky and gave a high-pitched howl. The hyena, its attention on the vixen, failed to notice that Jason, seemingly in response to Jewel's call, was also racing to the scene. A split second later the dog jackal bit the interloper's ankle. Again the hyena swivelled round and thus once more exposed his rump to Jewel, who darted in to nip the other ankle. Time and again the two jackals attacked the hyena, which finally lowered itself into a squatting position, thus protecting its ankles with its less sensitive bottom. In this attitude it slowly scuffled away, and as the jackals followed, darting forward in turn to nip its rump, the hyena pirouetted grotesquely, trying to defend both sides at once.

The incident reminded me of the observations of an American scientist who watched a family of wolves chase a grizzly bear from their den in a similar way, biting it in the rump. The hyena and the grizzly bear are somewhat cumbersome creatures and the jackal and the wolf are both capable of lightning movements – they can thus nip and get away before the larger animal has a chance of retaliating. This, presumably, is why they are able to chase off creatures weighing up to four times as much as they do themselves.

As the days went by my assistants and I saw Jason and Jewel chase off other hyenas, and eventually many hyenas made a detour as they passed, seemingly to avoid the trouble-spot. I have never actually seen a hyena eating jackal cubs – indeed, when hyenas passed the den in the absence of both jackal parents, they did no more than poke their heads down the hole and sniff the ground nearby. Yet a year later I saw a hyena dig some way into an occupied jackal den before wandering on, and I have several times watched hyenas chasing jackal cubs. I feel sure, therefore, that if food were really scarce a hyena would dig out and feed on jackal cubs so that there is undoubtedly good reason for the evolution of the parent jackal's harrying behaviour.

As the days went by Jewel became more used to a car standing near the den and she began to spend longer periods of time with her cubs: and when she was with them she was usually grooming them. Jason occasionally nibbled one for a while, just as he sometimes groomed his vixen, but we soon realised that Jewel was a fanatical groomer. She spent minutes biting gently at loose hair and dirt in her youngsters' fur, and if a cub got restless before she had finished with it and managed to wriggle away to join its siblings in play, Jewel often ran after it, bowled it over with a sweep of her paw, and then continued her work. Usually when she had finished with one cub she transferred her attentions to another. Sometimes, indeed, she joined in a game with her offspring, seemingly for the sole reason of manœuvring one of them away from the rest, pinning it to the ground and grooming it. Once she even ran a short way beside a cub and nibbled at it as it loped after a sibling.

All jackal mothers, of course, groom their cubs, though we have seen none of them do so as vigorously or frequently as Jewel. Grooming, in many creatures, has become an important affair socially and, as well as serving to cleanse the fur, may also help to strengthen affectionate ties within a group, or cement the relationship between males and females during courtship. It undoubtedly has this social implication in golden jackal society. Jason and Jewel groomed for longer periods of time when they were courting than at any time afterwards, and both parents groomed their cubs far more frequently than a mere cleansing of the fur would warrant. As the cubs grew older and developed their own distinct personalities, so their parents groomed them more often and for longer at a time.

Very brief grooming, or even isolated parts of the grooming pattern of an animal species, quite often appear during friendly interactions between two individuals, such as greeting ceremonies. When Jason and Jewel met after a separation they frequently approached and briefly nibbled each other around the muzzle or ears. And often, when a low-ranking adult golden jackal meets a superior, it will roll over on to its back, adopting the position with which a cub invites a parent or a sibling to groom it.

By the time Jason's four cubs were nearly a month old their dark subterranean colouring had changed to the pale hue of sunbleached grass – excellent camouflage for those youngsters born in the dry season, but of little use for the majority which are born, as these four, when the grass is green. By this time we were able to recognise the individual cubs – although, apart from Rufus who early got a tiny, almost invisible nick out of one ear, it is doubtful if we should have known them out of their family setting. Within the group, however, there were not only slight differences in physical appearance (one female was sandier in colour than her sister, one male had a longer muzzle than the other), but there were definite differences in their behaviour and personalities.

One female, Amba, rapidly developed into a groomer almost as enthusiastic as her mother. Constantly she pestered her siblings with her attentions, though frequently she was unlucky for, as they grew older, they became increasingly lively and spent much time in playing. Occasionally, when her mother was grooming a cub, Amba would join in as well, but she also liked to be groomed and she was as likely to push between Jewel and the cub being groomed – a manœuvre which was usually successful since the other one would seize the opportunity to escape from its mother and leave Amba the only available cub. Sometimes I even saw Amba push between her father and mother when the two were grooming: one of them usually groomed her as a result.

Cinda was the smallest of the cubs, the runt. She was by far the most timid, scuttling into the burrow at the slightest shadow cast by a passing bird or cloud, tumbling in head first when startled by a sudden clap of thunder. She was quick to show submission during play when a game between the cubs became rough and often, indeed, kept right out of her siblings' play, lying alone whilst they raced around her. Even so, she was often pounced on or knocked over by the stronger cubs.

Rufus was the opposite; from the start he was by far the most venturesome of the four, the first to wander more than a few yards from the den. Once, when he was still only about three weeks old, I watched as he crept towards six Egyptian geese which waddled past the den, each one about four times his size. The closer he crept, the faster the geese waddled as though reluctant to take to the air for so small a jackal. It seemed, too, that Rufus had an inquisitive nature: when a plant burst into flower near the den, Rufus must go and sniff at it; if a bird dropped a feather, Rufus was usually the first to notice and investigate. Amongst jackals, as amongst the young of many mammals, if one cub does something the others are likely to copy. When Rufus began pouncing on insects, the others did the same. When Rufus wandered five yards from the den to feast on steaming rhino dung, the others soon followed his example.

Nugget, the other male, always seemed the most playful of the four. Constantly he approached one or other of his siblings with the tiny head-shaking movements which mean 'I want to play with you'. When the others had tired of romping with him he played on his own, tugging at pieces of grass as though they were the tails of his siblings and then, as the stems broke, tumbling over backwards. Frequently, too, Nugget threw small stones or pieces of dry zebra dung into the air with his mouth and then pounced on them as they hit the ground. One day he found a chewed, dried-out ball of hair and skin, and this 'toy' he kept with him for over an hour, throwing it up and pouncing on it, laying it aside for a while but taking it with him when he moved away and playing with it again. Once I saw Nugget even shake his head as a play-invitation to a butterfly!

The play behaviour of the cubs changed as they grew older. The early gentle biting and tumbling games gradually gave way to quite vigorous biting, chasing and wrestling. Often, as two cubs played, each would have its front paws around the other's shoulders whilst the two bit at each other's faces. Then, as they lost balance, they tumbled over and frequently continued to play as they lay on the ground, snapping at each other and kicking out with their hind legs like cats. Often a game started by two cubs was joined by one or both of the others until there was a heap of golden fur, waving paws and snapping teeth.

Once we were familiar with the cubs as individuals their playing became more interesting to watch. One day Nugget found an ostrich feather. He sniffed it cautiously, then pounced on it and began to worry it as a domestic puppy worries a slipper. When Rufus came up Nugget picked up his feather and darted away:

Rufus immediately gave chase and soon their two sisters joined in as well. Round and round they raced until Rufus caught up and had a growling tug-of-war with Nugget. Suddenly Amba grabbed Nugget's tail and pulled, a favourite sport. Nugget, turning to bite at his sisters, lost the feather to Rufus. And so it went on for over an hour until only a few tattered shreds of their toy remained.

On another occasion I watched Rufus and Nugget chase each other round and round a patch of tall grass. Suddenly Rufus stopped, turned, and as Nugget appeared round the corner, going as fast as his short legs would take him, Rufus pounced straight in his face. Then Rufus was off again, but after running part way round the grass clump he stopped again, turning to wait for his brother as before. He never saw the little head craned over the grass, peering down at him, and Nugget jumped straight over the vegetation and landed on his brother's back. Now the two cubs reared up on their hind legs, biting at each other's faces. It was at this moment that Cinda arrived. She watched for a moment and then pounced, landing on the others and throwing them off balance. Rufus extracted himself from the bottom of the heap and at once started to attack Cinda, biting hard at her face, his ears back. Then Nugget, struggling out from under Cinda, bumped into his brother's head and caused Rufus to bite his own rump. With a squeal, Rufus ran away from Cinda, the runt: did he think that she had given him that painful nip?

This incident reminded me of the time when our son Grublin, then only six months old, was bitten by a tame mongoose. We were having tea with friends, and Jane sat down, with Grublin in her arms, on the sofa. Just as our hostess sat down beside them the mongoose jumped up, bit Grublin's foot and vanished under the furniture. Grublin never saw it and was obviously convinced that the bite came from the lady who had just sat beside him. He screamed every time she approached him for the rest of the afternoon.

*

As Jason and Jewel became more tolerant of our cars it became possible to follow them on their hunting forays, and we gradually learnt the extent of their home range, or hunting grounds, and the sort of relationships they had with some of the golden jackals who were their neighbours. So far as we could determine they hunted over an area of approximately one square mile – some golden jackals, as we shall see later, have much larger hunting grounds. Jason and Jewel did not behave consistently when they met other golden jackals in their hunting grounds: sometimes the intruders were chased off, sometimes ignored and sometimes approached in a friendly way. Only a long-term study could explain such differences in attitude, but we guessed that sometimes the 'strangers' were, in fact, relations. Occasionally, for instance, we saw Jason greet one member of a pair but ignore the other, whilst Jewel ignored them both: this is what one would expect if the one was Jason's sibling. At other times both Jason and Jewel greeted one of these intruders: perhaps it was one of their previous offspring now grown to adulthood.

However, even if neighbouring golden jackals *were* related to Jason, we soon found that they did not often penetrate far into his home range when hunting. If lions or hyenas made a kill well within Jason's hunting grounds he and his vixen were usually the only golden jackals feeding there. On the few occasions when one or a pair of 'stranger' adults were present, there was a good deal of fighting over the food. In the same way, Jason and Jewel normally kept away from kills made well within their neighbours' ranges. It seemed, however, that the different home ranges overlapped each other to some extent and kills made in such areas were frequently attended by two neighbouring pairs of jackals. Once we found Jason and Jewel feeding at a carcass with two neighbouring pairs when the area of the kill was roughly common to all three home ranges. Sometimes Jason and Jewel were tolerant of the other adults; sometimes there were threats and short fights. It depended which neighbours were involved.

There is a great difference between the home range of a jackal and his territory. The latter is a relatively small area within which he courts his female and raises his cubs. Jason's territory, so far as I could determine, consisted of a narrow strip of flat plain about half a mile in length and some hundred and fifty yards in width. I only once saw another golden jackal penetrate this area, and both Jason and Jewel chased it, fighting it fiercely until it managed to escape and run away.

Very often when either Jason or Jewel set off hunting they trotted for a while along one or other of their territorial boundaries, marking it with urine about once a minute. Both male and female golden jackals usually raise a leg when marking, but the female squats at the same time, raising one back foot only a couple of inches off the ground. After urinating both Jason and Jewel often made a few scraping movements with their hind feet, thus spreading the scent over a wider area. Occasionally the jackals marked tufts of grass with their dung, or scats, positioning this carefully right on top of the tuft. The jackal has two anal scent glands and, when he (or she) defecates, it seems that secretions from these glands may be squeezed out, thus providing an added declaration of identity.

The dens in which golden jackals rear their cubs are, of course, well within their carefully marked-out territories. It seems, too, that their howling ceremonies nearly always take place within their territories. During such a ceremony each member of the family, except for cubs under two or three weeks of age, raises its nose to the sky and howls, high-pitched and shrill. Usually they do not howl in unison but join in, one after the other.

When one family stops calling, a neighbouring family may take up the strain, and so it may go on, back and forth – possibly for miles. In the crater such ceremonies often took place in the evening and the early morning, and sometimes during the night. The ceremony, presumably, serves as an additional proclamation of territorial ownership, like the dawn chorus of the gibbons and the territorial songs of many species of birds.

When we arrived at the crater in January heavy rains had just started, and regular downpours kept the grass green and lush, providing the herbivores with an abundance of food. All around us as we drove through the grazing wildebeest herds there were cows heavy with young. The calving season began about a fortnight after we had started our study of Jason and Jewel. One day, when we had already seen a number of very young calves still wobbling on their dark wet legs, we came across a young wildebeest cow about to give birth. She was lying down when we drove up to her, and we saw the strong contraction of her belly muscles. Never have I seen a birth so fast, so smooth, so comical: afterwards I realised that this was almost certainly her first baby. As the calf flopped on to the ground, its head bursting through the gleaming transparent sac, the mother jumped straight into the air and, twisting in mid-leap, she landed on her knees in front of her newborn. If ever I have seen an animal stare in wide-eyed amazement it was then. When the calf moved the mother jumped back, again landing on her knees, not taking her eyes for a second off the strange creature which had so suddenly landed behind her. We feared, at the time, that she might abandon her youngster, for each time it moved towards her she jumped away as if terrified. Ultimately, however, after the calf had struggled for over half an hour to reach its mother's udders, she permitted it to suckle.

In fact, even an experienced cow usually prevents her newborn from suckling for the first few minutes after birth. By constantly moving her hindquarters away from the calf she forces it to keep moving; the more it moves the quicker it learns to use its unsteady legs, and the better chance it will have of escaping its predators. So effective is this that, almost incredibly, a new-born calf can run beside its mother, at a fair speed, three to ten minutes after emerging from the womb. Sometimes another wildebeest will approach and butt a new-born calf, often causing it to fall over, and then continue to butt when it gets up again until, eventually, both calf and mother run from him. Once Jane and I watched a whole group of wildebeests chasing after one new-born which, despite the fact that its legs splayed and moved in the strangest way, managed to keep on its feet during the chase. It seemed brutal, yet this too probably helped to ensure that the interval of complete helplessness following birth was as short as possible.

Another of nature's safeguards lies in the fact that a very high percentage of the wildebeest cows (and the females of many other herd-living creatures) drop their calves at almost exactly the same time of year. This means that the predators quickly become sated and, ultimately, ignore many of the youngsters.

Right at the beginning of the birth season, when I was following Jason on one of his hunting forays, I was amazed to see him rush at a new-born wildebeest calf that was lying beside its mother, and bite it in the leg. The mother was on her feet in a flash and charged Jason with lowered head. The jackal made no further attack on the calf and I was puzzled: Jason was a bold jackal indeed, but surely even he would not try and kill a wildebeest calf, at least three times heavier than himself, and lying beside its mother. It would be comparable to a European red fox tackling a domestic calf in a field with its mother – except that a wildebeest is more agile than a cow.

Two days later I got an inkling of what might have prompted Jason's strange behaviour. That day Jason and Jewel left the den to hunt together – the first time I had seen them do so since the cubs were born for, as I have said, the golden jackal is often a solitary hunter. After trotting for a while through the wildebeest herds, Jason suddenly stopped and sniffed the air. Then, with Jewel beside him, he moved purposefully towards a cow wildebeest which was standing with a tiny calf beside her. When the mother turned and threatened them with lowered horns, Jason and Jewel ran off a short way, but immediately turned and followed as the wildebeest and her calf moved away. Half an hour later the jackals were still keeping close to the pair and I was becoming increasingly puzzled.

All at once the mother wildebeest grew restive, lay down and delivered the afterbirth. I realised that, although the calf was dry, it could not have been more than a few hours old. Jason and Jewel, presumably by smell, had known that the placenta was still within the cow: their patience was at last rewarded.

This gave me a possible explanation of Jason's attack on the new-born calf. Perhaps the smell of the birth fluids was so strong that, for an instant, he had become confused and bitten the calf's leg in the belief that it was a lump of afterbirth.

During the next two weeks, the peak of the calving season, placentas provided an abundance of food for the jackals. Often when Jason and Jewel were lying near the den I would see them become suddenly alert, staring into the sky. Then I too would notice the tiny shape of a vulture, growing larger as it plummeted towards the ground. As soon as the jackals had ascertained in which direction the bird was going they streaked across the open plain, often arriving only seconds after the vulture itself and getting most of the afterbirth.

It was soon after the beginning of the wildebeest birth season, when Jason's cubs were just over four weeks old, that they began to eat solid food. Jackals, like wild dogs, feed their youngsters by regurgitating food; the first time we saw Jason do this he trotted over to the four cubs, deposited some meat on the ground and then ate part of it again himself whilst the cubs gathered round and fed with their father. Later the same day Jewel did the same. For the first few days the cubs, although they ate the new food without hesitation, often left some lying on the ground. At the time Jewel was still suckling them about three times during daylight hours, and the few night watches we did suggested that they nursed about twice during the darkness, though the moon, at that time of year, was so often obscured by clouds that we could not be sure.

As the days went by, and despite the fact that they were still drinking milk regularly, the cubs seemed to appreciate regurgitated food more and more, and soon would jump up when either parent returned to the den, wagging their tails and licking the adult's lips and face – behaviour which can be seen in an adult jackal when it greets a superior. Gradually it seemed to become more and more difficult for their parents to regurgitate – they would turn away from their exuberant offspring and open their mouths, but the cubs were there already jumping up and licking. Again the parents turned from them: it seemed that the cubs prevented the adults from getting their heads low enough to permit regurgitation to take place. Once I saw Jewel actually pirouetting round as she regurgitated, the meat spraying out of her like the sparks from a catherine wheel.

Sometimes, if a cub persisted in begging for more food after one of its parents had already regurgitated once or twice, the adult turned and bit at the youngster's nose, a common form of punishment in all three types of jackals as well as in wild dogs. Ironically it was often Cinda who got bitten in this way for, during the first regurgitation, it always seemed that by the time she managed to get close to the parent's nose most of the food had been eaten. Quite often she got nothing at all and had to content herself with licking the grass where the meat had lain.

We saw quite a number of wildebeest calves dropped close to Jason's den, and on these occasions the parent jackals often carried lumps of the afterbirth back to the cubs in their mouths. This was when the value of regurgitation as a method of feeding the young was brought home to us. Five times we saw tawny eagles swooping down over the parents when they had food in their mouths, and three times the birds actually hit the jackals' backs with their talons. On one of these occasions Jewel dropped her meat, seemingly in fright: before she could turn to threaten the eagle the latter swooped down and carried the food away. Another time Jewel dropped a large hunk of meat when she was chased by a hyena.

Sometimes, during these days, there was an over-abundance of food. Once, for instance, Jason returned to the den and regurgitated not once but four times. The cubs simply could not finish a four-course meal and several pieces of uneaten meat were left on the ground. Suddenly the cubs, with one accord, bolted for the den and, at the same moment, Jason leapt into the air, his teeth snapping, as an African black kite swooped down at the food. I was not surprised to see this, for these kites, though small, are audacious birds. In towns I have seen them snatch away bones that were actually being gnawed by dogs; and once, when Jane's mother was in a boat on Lake Victoria, a particularly daring kite dived down and grabbed at her head, apparently having mistaken her hair for some brown mammal. Four times Jason leapt up to bite at the kite before the bird gave up, and I was amused to notice that, after his first fright, Rufus cautiously poked his nose out of the den to watch what was going on.

Usually golden jackals bury excess food after the manner of foxes, wolves, bears and a number of other carnivores. Jason and Jewel always buried small pieces of meat separately, spreading their entire larder over a large area. I guessed why when I saw a hyena moving slowly and methodically from one small piece of food to another in one of Jason's larders: because the meat was so spread out it was far more likely that he would miss some of it. The fox, in Europe, has less to fear in this respect and may bury his whole larder in one place – a single cache was found which contained one hare, one grouse and ten mice.

The wildebeest birth season was the one time of year when we saw the golden jackals of the crater scavenge more than they hunted for their food: normally they scavenged only when larger predators happened to make a kill in their hunting grounds. It was on such occasions that I learned much about Jason's character, for never have I watched a golden jackal that was more aggressive or bolder at a kill. Once I saw him snatching mouthfuls of meat from under the elbow of a lion as it lay on its kill, and once he had a tug-of-war with a hyena over a piece of entrail. He leapt fearlessly at eagles and vultures feeding on a carcass, and one day, when his cubs were older, he ran for over half a mile to chase off a huge lappet-faced vulture that had driven his son, Rufus, from a piece of meat he was eating.

However, despite his efficiency in the role of scavenger, Jason far more frequently hunted and caught his own food. In fact, one of the pictures of Jason which springs most readily to my mind is that of a tense figure standing motionless by some tall clump of grass and then, as though an unseen hand had released a trigger, springing high into the air and landing, as often as not, on some unseen rodent which he had heard with his sharp ears.

My assistants and I followed Jason and Jewel on many of their hunting forays. Most frequently they caught insects, digging in the ground or in heaps of dung to uncover dung beetles or their larvae, termites and other such things; leaping up with a snapping of jaws to catch beetles or moths in mid-flight; pouncing on grasshoppers, crickets and the like as they moved through tall grass. Often, too, they caught rodents, such as rats and mice, and at some times of the year they preyed on ground-nesting birds and their young. Several times I saw them chasing spring hares at night, and once, in the morning, the cubs were playing with a piece of spring hare skin; whether or not their parents had caught it I could not tell. Occasionally we saw Jason and Jewel chasing an ordinary hare, but without success: more often we watched the jackals stop and watch when they flushed a hare and then trot on, as though they felt that pursuit would be useless. Even in areas where food is less plentiful than in the crater, and where golden jackals often chase hares, we never saw them either catch or eat one.

I shall always remember the first time I saw Jason fight a snake. I arrived on the scene just in time to see him leap backwards with the snake gripping his neck. Jason shook himself vigorously and the snake's fangs slipped from the jackal's long protective ruff. A split second later Jason had the snake in his jaws. He shook it briefly and then, again jumping backwards, let it go. The snake landed and, in a flash, coiled its body, its head pointing towards Jason. Suddenly it shot forward, its fangs gleaming briefly in the morning sunlight. But even as it struck so Jason leapt aside and, like lightning, grabbed hold of its body and shook it once more. After this Jason's attacks followed each other quickly until, after a few minutes, his victim lay almost motionless on the ground. Then, starting with the tail, Jason slowly ate the limp body.

At that time I believed that encounters between jackals and snakes would be rarely observed, but soon I found that snakes form a normal part of a jackal's diet. Once I saw Jason eating a snake without apparently killing it first, and I have since wondered whether jackals have an instinctive knowledge as to which snakes are deadly. His first victim was a striped sandsnake – only mildly venomous to humans but possibly more dangerous to a creature as small as a jackal.

There is another day I shall long remember. Jason and Jewel were wandering through a herd of Thompson's gazelle when Jason suddenly stopped and stared ahead. Jewel moved to stand beside him, and following their gaze I too saw the tiny fawn pressed flat against the ground. Nearby the mother grazed, as yet unaware that her offspring was in danger.

A moment later Jewel darted forward to grab at the ear of the fawn. But the mother gazelle moved as fast as the jackal and within seconds had charged to the fawn's defence, her head lowered, her small pointed horns directed towards the predator. Jewel was quick to dart away, but as the mother gazelle followed her, Jason ran in and seized the fawn. Swiftly the gazelle turned back and now she charged Jason. As Jewel had before him, Jason dropped the prey and raced from the mother gazelle. Time and time again the mother charged the jackals, first one and then the other, but even though she twice sent Jason hurtling head over heels, she apparently did not harm him. After eight minutes Jewel managed to run off with the prey and, presumably to escape the attentions of the mother gazelle, dragged it down a nearby burrow. Once her youngster was out of sight and no longer bleating, the gazelle moved away. Jason, who had not seen where his vixen went, stood looking around until, after a while, Jewel emerged with the dead body of the fawn. Soon she and Jason were feeding together on their prey.

During the next few weeks I saw several Thompson's gazelle fawns hunted successfully by both golden and silverback jackals: each time the jackals hunted in pairs. I also watched one hunt when the fawn managed to escape from a pair of silverbacks after being three times caught and actually carried by one of the hunters. Finally, whilst both jackals were being charged by its mother, the fawn ran amongst a herd of male gazelle nearby and pressed itself to the ground. After this the jackals sniffed about for a while, but seemed to have lost the scent and finally wandered off.

The diet of an individual jackal seems to depend, in part, on where he lives. Thus some of Jason's neighbours, who hunted in hillier country where the grass was long, had more chance of hunting birds, snakes and even rodents than Jason and Jewel, whose hunting grounds mostly comprised short grass plains. The range of some jackals included the banks of the Munge River and, when the figs were ripe, they feasted on fallen fruit – a delicacy which, so far as I know, Jason and his family were denied. Indeed, there were few fruits to be found in Jason's hunting grounds: if these are available many kinds are eagerly eaten by jackals. At certain seasons of the year, however, Jason's home range sprouted a variety of mushrooms less common in the high grassland of his neighbours' hunting grounds. Several types of these were eaten by Jason and his family. I shall never forget when the cub, Rufus, ate a mushroom of a type I had not seen jackals eat before. Ten minutes later he seemed to go mad. He rushed around in circles and then charged, flat out, first at a Thompson's gazelle and then at a bull wildebeest. Both animals, possibly as surprised as I was, hurried out of his way. Could the mushroom have caused hallucinations? Had Rufus been on a trip? The question must remain unanswered as I could not find another for identification.

It seems, therefore, that the jackal enjoys a balanced diet, throughout much of the year, of meat, insects and sometimes fruit. We do not know, as yet, to what extent the feeding patterns of jackal cubs are moulded by the sort of foods brought to them by their parents. I am often asked whether jackal cubs learn to hunt by watching their parents. So far as some foods are concerned this is certainly not so. Rufus and his siblings first started to hunt insects, digging at the grass and making snapping movements with their jaws, when they were only about three weeks old. They responded, presumably, to the movement of the insects. At that age the hunting movements were either playful or ineffective for we seldom saw them eating insects until they were a month old. Later still, after they had been weaned, the cubs took up insect hunting in earnest.

Jewel started to wean her offspring when they were just over two months old. Sometimes she allowed them to suckle as usual, but at other times she jerked her hind body away. If they persisted, she hurried away, sometimes tripping over one or two cubs as she went. Occasionally the cubs tried many times to reach their mother's nipples, and although Jewel sometimes punished them with a soft nose bite, she more often groomed the offenders: a strategy also practised by chimpanzee and some monkey mothers and one which usually serves temporarily to distract the youngsters from their desire for milk.

After they had been weaned it seemed that the cubs suddenly became more independent. Led usually by Rufus, they moved up to twenty or more yards from the den whilst hunting insects or simply investigating their surroundings. On one such occasion Nugget discovered a frog. He pounced towards it but, just before he landed, the frog leapt away. Nugget jumped again, and I laughed out loud as I watched these two very different creatures bounding along one behind the other. When the frog finally vanished into a pool of rainwater, Nugget stopped dead and, after a moment, I saw him staring intently at his image in the puddle. Cautiously he put his nose towards the nose of his reflection, jumping back as he touched the water, sneezing and shaking his head. A moment later he cautiously touched noses with his image again, and once more jumped away. As he was returning for the third time Rufus joined him and the brothers stared, for a moment, at their two reflections. Then, with an investigatory touch of his paw, Rufus shattered the images.

The period of weaning also corresponded closely with the first signs of real aggression between the cubs. Sometimes, during play, Amba would bite hard enough to draw a squeal from Cinda. At other times Rufus, after buffeting one of his siblings to the ground, stood over the victim looking just like an aggressive adult, the hair of his neck and shoulders erect, his tail held stiffly behind him. Growls and snarls crept into the play vocabularies of all the cubs.

Rufus was the first of the cubs to incorporate 'body-slamming' into his games. This behaviour is seen in adult golden jackals in aggressive situations – particularly when they are threatening vultures or eagles at a kill. The aggressor faces its opponent and then, often leaping with all four feet off the ground, swings its body round through 180° to slam its hindquarters into the other. It is a particularly good way of chasing off vultures, for not only does it intimidate but it protects the jackal's eyes from the claws and beak of its opponent. Eventually all the cubs used body-slamming during play, along with play-biting and play-wrestling, but always it was Rufus who was most likely to bully, transforming a game into fighting at the least provocation. And it was Cinda, of course, who most frequently fell victim to the aggression of the others, for she was the quickest to become submissive. She would crouch and cower in response to behaviour which the others would probably have interpreted as play. Cinda kept more and more to herself, frequently lying alone whilst her siblings played together.

Probably it is during the play of cubhood that jackals, like many other creatures, start to sort out the social position they will hold in later life – at least in relation to each other. It really seemed, as we watched the four cubs, that from two months onwards each one, apart from Cinda, wanted to establish its superiority over the others. Nugget and Amba were pretty evenly matched, though Nugget usually managed to get the better of his sister in the end. But it was obvious, from the start, that Rufus, the strongest and the most aggressive, would always dominate the others.

One day when the cubs were about ten weeks old Jason trotted back to the den with the hindquarters of a new-born Thompson's gazelle fawn in his mouth. This was the first time we had seen the cubs fed with a really large hunk of meat, complete with skin and bones. Almost before Jason had let go of the 'kill' it was appropriated by Rufus. Nugget and Amba rushed up and tried to join in, but Rufus turned with ferocious snarls, snapping at their noses, and they backed away. I was reminded of the transformation which took place in my tame bat-eared fox every time I fed him a mouse, or other small creature, complete with its skin. He became, in an instant, a fierce wild animal, and it was a foolish person who tried to touch him then. So it was with Rufus. Amba quickly gave up trying to share the food, but Nugget persisted: each time Rufus threatened him he, in turn, snapped at Amba, directing towards her the aggression he dared not show to his brother.

Cinda kept well away from all this and so, when Jewel arrived an hour later with the head of the fawn in her mouth, only Cinda noticed. For once the runt feasted unmolested. And she was still feasting after another hour had passed when Rufus, his stomach bulging, slowly moved away from the remains of his gigantic meal. He had left only bones, some tatters of skin and the two little hoofs for Nugget and Amba. Suddenly Rufus stood stock still and stared towards Cinda. Then he waddled over as fast as he could and, as she cringed away from him, quickly relinquishing the head, he picked it up and staggered away with it. He carried it some thirty yards from the den and buried it in a large heap of zebra dung. Then, with the self-satisfied air of one who has done well, he flopped down to sleep. I couldn't help chuckling at his dog-in-the-manger tactics, particularly as Cinda, despite him, had managed to enjoy an hour's uninterrupted feeding.

In other ways, too, the cubs were growing up. Sometimes when Jason or Jewel left the den to go hunting the cubs would follow, at first only for twenty yards or so but, as the weeks went by, for longer and longer distances. We noticed, too, that the youngsters were starting to sleep out on the grass more often than in the darkness of their den. One day as Nugget lay fast asleep in a clump of grass a young wildebeest started to eat away his cover. For a minute nothing happened, but when it seemed that the calf would soon be nibbling at Nugget's straw-coloured fur he suddenly raised his head. The wildebeest, pausing in mid-chew, stared for a second; then the two fled in opposite directions. I was amazed then, as I have been so often since, at the incredible lack of alertness shown by sleeping jackal cubs – it must, I am sure, account for a number of deaths.

The first disaster to strike the Jason family occurred when the cubs were about ten weeks old. Jason was curled up near the den and the cubs were spread out insect-hunting nearby. It was a cloudy day with no sun to throw a warning shadow, and none of us saw the black shape silhouetted against the grey sky until it had half folded its wings to dive. Then we heard the air whistle through its feathers as it plummeted to the ground. For a split second the jackals froze, but when they heard Cinda's terrible scream as the eagle's claws gripped her they ran – the other cubs towards the den and Jason towards the eagle.

Slowly the bird, a bateleur eagle, carried the screaming Cinda off the ground whilst Jason ran along underneath, his head pointed up as he watched the drama above, unable to help. The bateleur is one of the smaller eagles and it had difficulty in gaining height with its comparatively heavy burden. Suddenly it let go of its prey and Cinda hurtled to the ground. I felt certain that she would survive neither the wounds made by the eagle's talons nor the impact of landing – for she had fallen some twenty feet.

Jason at once ran to the place where Cinda had landed, in a patch of high grass. I drove over there too when Jason looked calmer, but I saw no sign of Cinda. For the remainder of the day I watched the rest of the family without enthusiasm. For the most part the cubs lay around the entrance of the den, and twice they darted down when birds flew overhead. Jason went off hunting and, eventually, the sun sank. As I lay waiting for sleep that night I could still hear the whistling of wind through folded wings and Cinda's terrible scream; and I could still see the small golden body crashing down from the sky.

At the den next day life continued as if nothing had happened. Jewel slept in a clump of grass, Rufus hunted insects and Nugget played with a stone, holding it with his front feet as he lay on his side and pawing it with his back feet like a cat that plays with a ball of wool. Then he played with Amba when she came over to try and groom him. Jason was not visible, and I presumed he was out hunting.

Two hours after arriving at the den I started as a small figure slowly appeared from the den. It was Cinda. She walked stiffly, and when I examined her closely through my binoculars I saw she had a deep cut under her chin. I could see no other visible injuries. She blinked in the sunlight and then lay close to the den entrance. After a while Amba groomed her.

Cinda's wound turned into a nasty abscess and for nearly a week she was lethargic. Jewel and Amba frequently licked the place, however, and Cinda's health gradually improved until, three weeks later, she was back to her normal self.

Cinda, in fact, was lucky. A few weeks later I saw a cub from a neighbouring jackal family taken by a Martial eagle – a huge bird of prey known to carry off young gazelle, monkeys and other such victims. That cub did not survive. Later that year one of our assistants saw another Martial eagle trying to fly off with an almost full-grown silverback jackal. As it gained height, however, it was attacked by a large vulture (probably a lappet-faced vulture) and, eventually, was forced to land. For a few moments it tried to beat the vulture off whilst dragging its screaming captive along the ground, but eventually let go and the jackal managed to creep away. Birds of prey, in fact, probably constitute one of the major threats to young jackals: Jane and I found the remains of three jackals in and under a tree in which there was a large eagle or vulture nest. The only other predator known to prey quite often on jackals is the leopard.

Once I had seen an eagle make its deadly attack I never again felt quite at ease when Jason's cubs were playing or hunting far from the den. I was particularly apprehensive when they made their periodic moves from one den to another, which happened five times before they were twelve weeks old. The longest distance between the first four dens was only seventy yards, but the fifth den was half a mile from the fourth. This last move was the only one that we were able to watch as it took place in daylight. That morning, soon after I arrived at the den, Jason and Jewel, who had been lying nearby, got up and trotted briskly away. Three of the cubs left their insect-hunting and followed, but Cinda remained where she was, sitting at the entrance of her den. It seems that some signal must be given by the parent jackals to indicate to their cubs either that they are allowed to follow or that they may not – why else should the three cubs have gone on without hesitation whilst normally they turned back after about twenty yards? Anyway, they followed Jason and Jewel until some high vegetation hid them from sight. I drove after them then, leaving Cinda behind, a tiny figure silhouetted against the green dew-spangled grass.

The parent jackals never hesitated as they trotted along. From time to time one or more of the cubs bounded past them or lagged behind to investigate something, but on the whole the family

1. A jackal pair hunting in the Ngorongoro Crater.

2. The cubs rush up to greet their mother, Jewel, when she returns.

3. The cubs spent much of their time at play.

4. Nugget pounces on Rufus as he hunts insects.

5. A tug-of-war was a common game.

6. Rufus with a wildebeest horn is pursued by Nugget and Amba.

7. Jason pounces on a rodent.

8. Jason in a tug-of-war over some meat with a hyena.

9. Jason in a battle with a side-striped sand snake.

10. Much of the jackal's (in this case silverback) scavenging takes place at night.

11, 12, 13. A jackal catches a Thompson's gazelle fawn, but lets go when the mother charges.

14, 15. A dominant silverback jackal often throws his body sideways at a subordinate, thus slamming his rump against his opponent and sometimes ending with a backward kick.

16. A silverback jackal defends its meat against a marabou stork.

17. Nugget and Amba howl in response to their parents' calls.

18. Cinda watches as her two brothers play.

19. A two month old jackal cub plays with its one year old sister.

20. As the cubs grew older, they were more often aggressive towards each other.

21, 22. The male of a courting couple does not tolerate possible competition.

23. Nugget, even though almost fully grown, was still submissive to his mother.

24. Jason and Jewel grooming Amba.

25. Howling served to establish contact between parents and the cubs.

stayed close together. When we had travelled about quarter of a mile from the den the small procession vanished into a patch of tall grass; when it reappeared I noticed that Amba was missing. I waited for a while, but she did not show up and so I drove on after the others, none of whom turned to look back or even slackened pace.

Why had Amba dropped out? Perhaps she had got scared so far from the familiar landmarks of her home. Presumably she could find her way back as the fresh scent would lie heavily on the dew-covered ground. But I wondered about the dangers she might meet; would she be able to escape if she came face to face with a hyena or if an eagle swooped down? It seemed unlikely. True, there were holes into which she could run, but what might she meet in their gloomy depths? I should like to have searched for her, but I was afraid she might panic if I approached when she was alone and far from her den and so lose her way. So I drove on with my fingers metaphorically crossed.

After journeying for another quarter of a mile, Jason and Jewel lay down, whilst Rufus and Nugget eagerly investigated their new surroundings. Soon they had both discovered the new den to which their parents had led them. There was freshly dug earth at the entrance, and I wondered whether Jewel had been preparing the new burrow for her cubs during the previous night.

Later in the day first Jewel and then Jason went off hunting. Rufus and Nugget seemed quite at home in their new surroundings and presently I left them there and drove back to the fourth den. To my relief I found both Cinda and Amba there, curled up and sleeping in the hot sun.

For the next few days Jason and Jewel looked after the cubs at both dens but, on the fourth morning after the move, I found all four cubs reunited at the new den.

Watching one jackal family is a full-time job, for if you have to miss a day you feel convinced something tremendously significant will happen in your absence. And so, but for the untiring help of the students, particularly Ben Gray, who helped me for so long with the jackal study, I should have been able to learn little about other jackal families. As it was, one of us frequently spent several days or weeks watching jackals in the surrounding country, and I discovered that I was lucky indeed to have picked Jason and his family for study. In other families it seemed unusual for as many as four cubs to survive beyond the first few months. One of Jason's neighbours was first seen with three cubs; one was taken by an eagle and another disappeared, leaving but one to grow to adulthood. Of another family comprising two adults and six cubs, two cubs vanished and a third followed his father on a two-mile hunting foray and was left by the male jackal near the Munge stream. The father returned at dusk to the place where he had left the cub eight hours earlier, but the youngster had gone and I never saw it again. At another den one of a pair of cubs disappeared during the night; there had been a hyena kill near the spot and much jackal howling during the hours of darkness, and we suspected that the hyenas were responsible for the cub's death. And a heavy rainstorm that flooded much of the crater basin may have drowned two tiny cubs, for after the level of water had subsided, and we were able to reach that den again, we found only two instead of four youngsters.

This flooding occurred when the rains, which were unusually violent that year, reached their climax, and we too were affected. It happened one night. It was a beautiful night, brilliant with stars and there were no rain clouds, no rumbling of thunder, to warn us of impending disaster. It was I who discovered the flood for I walked behind the tent and found myself knee-deep in water. I shouted for help and light, and as Parker and Ben Gray, our two American students, and the two Africans, brought lamps from the dining tent and kitchen which were on higher ground, we stared around us in dismay. The cabin, also on a slight rise, was surrounded by water and between it and the river bed the flood waters were already flowing into our three small tents. Quickly we started to rescue bedding and clothes. The beam of a torch briefly illuminated a pair of male underpants before they were whirled on into inky blackness. There was an anguished yell as someone, in bare feet, trod on some virulent African stinging nettles under the water and, doubling up, dropped a pile of bedding. A sudden wail from

Grublin woken from his sleep as more and more sodden things were thrust into the cabin. A horrified American voice calling 'Oh, my God!' out of the darkness, and then, a few seconds later, 'Shucks! it's only a frog'.

And still the water was rising as we struggled together to save the tents. The first two were not difficult but the third, with a sewn-in groundsheet like the others, was on lower ground. There was no outlet for the water at the back and I was up to my waist in the flood as I tried to free the bending metal tent poles. Suddenly there was a tearing sound as a pole forced its way through the canvas and, as the tent collapsed on top of me, I felt the whole thing beginning to surge away in the darkness. I panicked for a moment as the heavy canvas pressed me down into the water, but Jane, Parker, Ben and the two Africans soon had things under control and we managed to salvage the tent, although we lost several tent poles and most of the pegs.

It was midnight when we finally sat down, exhausted and soaked to the skin, for a cup of coffee. Then we realised that the night was strangely quiet. The rushing of the stream as it poured over little waterfalls and gurgled through the roots of the fig trees had given way to a sound like a soft breeze as the waters flowed through high grass.

After these last violent rains of the wet season the floods gradually subsided and we no longer returned home, almost every day, covered in mud from jacking our cars out of one pot hole after another. Jason's cubs were nearly four months old by now and they spent less and less time down in their den. True, they normally hurried towards it at the approach of a hyena, but they often turned around on reaching the entrance and watched when one or both of their parents attacked the intruder. They seemed fascinated by the sight of the hyena pirouetting away from the sharp teeth that so unerringly nipped its ankles or bottom.

One day when Jason and Jewel were out hunting I watched as an old female hyena ponderously approached the cubs. I recognised her as Mrs Brown, for by then we had started our hyena study. Rufus, Nugget and Amba had curled up together, Cinda, as usual, lay by herself. Slowly, sniffing along the ground, Mrs Brown got closer. Then she paused and looked towards the cubs, her mouth half open in the heat and a long string of saliva drooling to the ground. Just as I was wondering whether she would see or smell the sleeping jackals, Cinda started flicking her ears. Unaware of danger, she sat up, snapped at the insect bothering her, and then lay down and closed her eyes. As Mrs Brown started to move on, Cinda again began to flick her ears, the movement clearly revealing her whereabouts. Mrs Brown got closer. Cinda snapped once more at the insect and then lay motionless.

There was a howl from Rufus, a sudden movement, as the hyena rushed forward with jaws snapping at Cinda's slender body. But Rufus's warning had come just in time and Cinda streaked away, a golden ball bounding through the grass. Mrs Brown lumbered after her for a few yards, but quickly gave up the chase, for Cinda's speed now matched that of her parents.

A few days later we again saw hyenas approach the jackal cubs, but this time the jackals saw them coming and, instead of running away, they approached within a few yards of the visitors and, lifting their noses into the air, gave the adult alarm call. I was interested, on that occasion, to see that Jewel, who must have been lying or hunting nearby, came running towards the cubs in response to their calling but stopped some distance away and, after surveying the scene, turned and went off. She seemed to realise that her offspring could tackle the situation without her help.

During our last weeks at the crater we noticed that the jackal cubs, nearly half-grown in size, were spending less and less time interacting with each other; if they played together the action was usually rough and often ended in fighting. Sometimes, when two began to squabble, one or both of the remaining cubs would join in the attack on the subordinate – on such occasions the den still proved a good bolt hole for the victim. Often, of course, the victim was Cinda. For the most part we found each cub hunting on its own, fifty yards or more from its nearest sibling, and often they rested apart as well. Jason and Jewel, although they were still feeding the cubs, spent less and less time near the den, and I felt sure that the family, which had seemed so united, was slowly splitting up.

*

Finally the day came when we had to leave for, much as I should have liked to continue the study on a long-term basis, time was pressing and I had to start collecting information on the wild dogs. I planned, however, to return to the crater for a couple of short visits during the following eighteen months to check up on Jason and his family. In fact, we were able to learn quite a bit more about golden jackals whilst we were camped at Lake Legaja; it was especially interesting to compare certain aspects of the behaviour of the jackals of the open plains with those living in the crater.

One difference lies in the size of the hunting grounds. In the crater the largest I knew of was about two square miles, whilst many of the jackals there, Jason included, ranged over areas half that size. On the short grass plains of the Serengeti, however, most pairs of golden jackals appeared to hunt over four to nine square miles. The reason for this, I think, is fairly clear. In the dry season these plains, with all surface water dried up, appear an inhospitable environment indeed, and the jackals living there undoubtedly need a large area to hunt over in order to find enough food. Indeed, I watched one, at the height of the dry season, trot along for almost a mile, pausing occasionally to dig at the ground or listen, without finding a single thing to eat.

Nevertheless, the jackals we saw there seemed healthy even when the conditions looked at their worst. We examined as many of their scats as we could and found that they were eating a number of insects, particularly dung beetles and their larvae, lizards, rodents and an occasional snake. During some months their scats were comprised, almost entirely, of the remains of various types of fruit. As I got to know some of these jackals better I became increasingly convinced that the species can survive for long periods without water, although, since I was not making a detailed study, I could not be completely sure that they never travelled long distances in order to drink.

I soon realised that it was more common on the short grass plains to find four, five or even six adult golden jackals around a kill than it had been at the crater. The fact that food is scarce for much of the year probably gives these plains jackals a greater incentive to penetrate each other's hunting grounds more deeply and more frequently when they get wind of a kill than is provided by the more hospitable environment of Ngorongoro. On one occasion Jane and I drove up to a dead zebra and were amazed to see no less than fourteen seemingly adult golden jackals gathered around it. Closer scrutiny, however, showed that only six were fully mature – the others were smaller and of slighter build and were, almost certainly, the nearly grown cubs of the adults. We were interested to see that, even with this many jackals around, there were seldom more than six or seven feeding together at any one time, and usually not more than three of the fully mature individuals. A number of quite fierce fights broke out, from time to time, between the grown males.

As I have said, I have not made a detailed study of the golden jackals around Lake Legaja, but nevertheless I came to know a number of pairs quite well, and it seemed that they, at any rate, stayed put in one area throughout the year like those of the Ngorongoro Crater. I have not yet found out what happens to their grown cubs when they finally leave their parents, nor do I know whether these youngsters ever follow the migratory herds of wildebeests, zebras and gazelles. We did, however, find out something about the habits of the silverback jackal in this respect. When the migration was concentrated around our camp at Lake Legaja we saw silverbacks, often in groups of six or more, wherever we drove. But when the migration moved on most of these jackals moved on too, leaving only a few resident pairs, which we came to know well, to hunt the bush area around the lake.

It was obvious, therefore, that many silverbacks become nomadic for some months of the year, following the migration on at least part of its annual trek. Usually these nomads seemed to be unpaired adults or young adults of both sexes: possibly, as is the case with lions, those silverbacks with established territories and hunting grounds never leave them, whilst those who are too young or too weak to establish their own territories follow the migrating herbivores.

Jane and I were especially interested to watch something of the social behaviour of these nomads. Presumably, when on the move, such a jackal occasionally meets other nomads of his kind whom he has never met before, as well as many whom he knows but slightly. In addition, he must sometimes pass through the hunting grounds of resident silverbacks. Would he try to avoid encounters with strangers of his kind, we wondered; if he was forced into proximity with them, at a kill, would there be much fighting as each tried to determine its social status?

We watched the behaviour of silverback jackals only long enough to whet our appetites – they will make a fascinating study. I remember clearly the first time Jane and I saw a meeting between two of these migratory jackals. The rather pale silvery-coloured male we were watching suddenly sat up, his ears pressed close against his head, his mouth wide open. Looking around, we saw that a second boldly-marked male was approaching at a brisk trot, his tail held out horizontally, his ears pricked. His whole attitude suggested that of a normally high-ranking individual. Maybe our jackal had met him before; if not, then the self-confident bearing was sufficient to convince him of the other's superiority. As the obviously dominant one approached, the submissive male raised one front paw high in the air and, when the other stopped close by, touched him lightly on the shoulder as though warding him off. For a moment the dominant male stood motionless and then, with lightning movements, he swung his rump through 180° to crash into the other's body. Twice more in quick succession he repeated this body-slamming, ending with a final backward kick which caught the crouching subordinate on the shoulder.

After this the dominant jackal trotted away and vanished behind a bush, reappearing a few moments later with a small piece of dry dung in his mouth. This he laid on the ground in front of the other male. We were reminded of the ritualised gifts which some birds and spiders give the opposite sex during courtship, and were mystified. But after a few moments, as the subordinate jackal made no move to accept the dung, the other picked it up again and threw it high in the air with his mouth. As it landed he jumped towards it and hit it forward with his nose. Then once more he threw it up and this time caught it as it fell. Now at last the subordinate male got up and for the next half-hour we watched as they played, chasing round and round a bush, tugging at opposite ends of twigs, jumping on to each other from a fallen tree. The offering of dung had been an invitation to play.

As the days went by we saw these jackals meet time and time again; nearly always their greetings included body-slamming, and often they were followed by play. Frequently, too, we watched meetings between two groups of silverbacks; often after various greetings between the different individuals they played happily together.

The greeting ceremony, with its harmless form of aggression, is not difficult to understand, for undoubtedly it enables stranger jackals to find out which is the strongest or boldest, by bluff rather than battle. If they have met before the greeting will serve to remind the subordinate of his rank. In either case the ceremony helps to reduce the risk of serious fights which might result in either or both participants becoming crippled.

The frequency and vigour of the play sessions amongst the silverbacks was more puzzling. Previously we had seen adult jackals play but rarely: sometimes parents (both silverback and golden) play briefly with their cubs; more frequently young adults engage in rather wild play sessions. But the play we saw between the nomadic silverbacks sometimes involved as many as six fully mature jackals and lasted up to half an hour. Why should this suddenly happen – why should normally non-playful adults suddenly engage in the tumbling games of youth? Was it, perhaps, the super-abundance of food which surrounded them at the time, taking them back, as it were, to the carefree days of their cubhood? This, undoubtedly, is part of the answer. Over and above this, however, a play session which follows a greeting ceremony probably serves to reinforce each jackal's awareness of his relative social position in a group of animals with which he is unfamiliar, or relatively unfamiliar.' And the better he knows this, the easier it will be for him when, later, he meets the others around a kill: he can take good care not to thwart his superiors and so run the risk of being attacked.

At the Ngorongoro Crater we have not seen adult silverbacks play together, nor have we often seen the spectacular greeting ceremony with body-slamming when they meet each other. At the crater, as on the plains, fifteen or more may gather round a single kill, but, so far as we know, these silverbacks do not migrate and thus the adults are presumably well acquainted. Each knows his status in relation to the others and so, it seems, the need for ceremony and play does not arise.

Perhaps our most interesting observations on the golden jackals of the open plains concerned their hunting behaviour, and it was something we first saw when the migration of wildebeests and zebras had moved out of the area leaving the thousands of Thompson's and Grant's gazelles. I was scanning the plains when suddenly my attention was riveted as I saw, through my binoculars, four creatures over half a mile away race over the plain. The heat haze deformed them so that they looked like ghosts flying through a mirage, yet I felt convinced that I was watching three jackals chasing an adult Thompson's gazelle. The gazelle appeared to turn and threaten the jackal behind it before running on once more. A few moments later the quarry again turned and ran at the nearest jackal which jumped aside. Then the other hunters caught up and a moment later the four shapes seemed to disappear into the ground. The distance was so far and the haze so bad that I could not be sure of what I had seen. I had already started the car, but it was fully two minutes before I had found a landmark in that vast open space so that I could be reasonably certain of arriving near the right spot.

When I got there the scene was no different from that which we had left. Around us the gazelles grazed and, with the approach of evening, some were playing, chasing each other in zig-zagging circles. After looking around for a while I felt sure I had imagined everything, but decided to drive about a bit to make sure. I had moved no more than twenty yards when suddenly a face covered in blood seemed to emerge briefly from the ground and then disappear. Quickly I headed towards the place and there, behind a small rise in the ground, were the three golden jackals eating an adult female gazelle. As one of the jackals bit into her flesh the gazelle gave a feeble kick, but after that she moved no more.

At that time, so far as we know, no other scientists knew positively that jackals were able to hunt down adult Thompson's gazelle, though the question was one which we had often heard discussed. It was exciting indeed to obtain first-hand information, and for the next few days I interrupted my wild dog study to try and learn more about this behaviour.

Time and again we watched hunts when the quarry escaped, and twice we drove up just as an adult gazelle gave its final struggle, but we never saw the actual moment of capture. However, as there were never any marks on the nose or throat of a victim, it is safe to assume that the jackal kills, like the wild dog, the hyena and the wolf, by disembowelling its prey. I was interested to learn that the coyote, so close to the jackal in many ways, goes for the throat when it hunts ailing adult white-tailed deer in America.

I soon realised that when the jackals set out to hunt adult gazelles they nearly always moved in groups of three to seven. There was one occasion though when I watched a lone jackal attempt such a hunt. It chased a female gazelle for over two miles, after which both hunter and hunted, panting heavily, had slowed down to little more than a fast trot. The gazelle ultimately moved into a large herd of its own kind and the jackal, possibly having lost sight of its quarry, or possibly through exhaustion, gave up.

On the whole, though, it seemed that these jackals worked, like wolves and wild dogs, in 'packs'. It took me a while to realise that these relatively large groups of jackals, so unlike anything I had seen in the crater, were not, in fact, completely adult groups; two individuals were nearly always a full grown male and female, whilst the remainder were slightly smaller and, almost certainly, the grown offspring of the adult pair. Only a long-term study of these plains jackals can prove whether this is really so.

*

Four months after we had left Jason and his family, Jane and I were able to make the first of our short check-up visits to Ngorongoro Crater. As we drove down the steep track from the rim the basin below looked dry and empty but, as usual, we found more wildebeests, zebras and gazelles grazing the brown grass than we had expected. We arrived at the Munge cabin in the afternoon and quickly unpacked the car. Compared with our previous large camp this one took no time to set up and, taking Grublin with us, we were soon out looking for the jackals.

It was exciting to drive over the familiar track, but the country looked as dry and barren as the Serengeti plains we had just left. Stretched before us was a semi-desert, the remaining bits of dry grass fighting a seemingly hopeless battle with the dust which twisted and twirled into the air with every puff of wind and then fell, powdering heavily all that lay below. When we finally reached the area where we had left the cubs all seemed derelict and lifeless. A cobweb, torn and dusty, clung to the entrance of the fifth den. Nearby lay a wildebeest skeleton, its skin clinging in places to the dried bones. The beard flapped in the wind as if, like some scarecrow, it would frighten all life from its surroundings, defending the green grass that must one day come to bring back its companions.

We only had a couple of hours in which to search and so it was not really surprising that we found neither Jason nor any of his family, but I was depressed, all the same. The next morning I set out alone. As I approached the area where Jason's cubs had been born I saw, to my huge relief, four adult jackals moving about. None of them showed the slightest sign of alarm as I drove up and I knew they must be part of the Jason family. In fact, I recognised Jason himself straight away: the other three were younger. Before I could sort them out a fifth jackal trotted up and lowered its head to the entrance of the burrow.

I could hardly believe it when five tiny cubs tumbled out of the entrance, quite as small as Rufus and his siblings had been when first I knew them. As these cubs began to suckle I realised that their mother was Jewel, and I was amazed that she should have given birth to another litter only six months after weaning the previous cubs.

As the morning wore on I gradually sorted out the other jackals, and I checked my identifications against close-up photographs that I had taken before leaving the crater: for we had discovered that each jackal has its own distinctive pattern of whisker growth. Amba, Cinda and Nugget were the three adult offspring at the den. Nugget had grown into a fine-looking young male, but when Rufus trotted up later in the day he quite overshadowed his brother: nowhere had I seen a more splendid jackal. His coat was a dark golden hue flecked with ruddy brown and the ruff around his neck was long and thick.

By the end of the day I had the impression that the four grown offspring were far more friendly towards each other than they had been when I left them four months before. Rufus showed no signs of aggression towards his siblings – presumably having established his position there was no further need for him to be a bully. Cinda had changed too, for she seemed to have lost much of her previous timidity and did not hesitate to join her siblings in a somewhat boisterous game.

It was Amba who spent the most time with the new cubs; indeed, she seemed unable to resist them. She groomed them incessantly when they were above ground, and once I saw her pick one of them completely off the ground by its hair and then, having put it back on the grass, keep it there by putting both front paws firmly on its back whilst she continued to tug its short fur with her teeth. On another occasion Amba noticed that a cub was slightly constipated. She hurried over and removed the hard pellet with her teeth, dropping it to the ground as the cub, with obvious relief, wobbled back to the den.

I was amused when Nugget approached his diminutive siblings and made playful head-shaking movements towards them. But the cubs were not old enough to respond and Nugget seemed unsure as to whether he should play with them or not. Finally he lay down close by and watched them, his ears pricked and his head slightly on one side.

When I arrived the following morning an amazing sight met my eyes. Some fifteen yards from the den where I had left them was a little row of jackals – Jason, Jewel and Amba with four tiny wobbling cubs. They were obviously heading for one of the other dens and still had some ten yards to go. I never imagined that such tiny cubs would travel on their own feet; they were so unsteady that time and time again they were thrown off balance by mere strands of dry grass. Twice, when I drove too close, Jewel got worried and put her mouth around the neck of one cub as though to carry it, but when I stopped the car she was reassured and left the youngster to continue its wobbling journey on its own feet. The den was reached without mishap and the cubs tottered into their new home. When they emerged half an hour later they still numbered only four. I presumed that the fifth had remained at the other den and would be moved later, but I never caught a glimpse of it despite the fact that I could see the old den from my place near the new. Nor did I see Jason or Jewel go near the place.

The next day there were only three small cubs wobbling around near the den and I had a premonition that all was not well. The following morning when I found only two cubs sitting near the entrance of their burrow I was even more alarmed. It may have been my imagination, but I thought they seemed more lethargic than before. I felt depressed, although I still hoped, without really believing, that the other three cubs might be in another den.

I arrived at the burrow the next morning just as the sun began to appear over the rim of the crater wall. Its orange glow tinted the early morning mist which hung just above the ground, for it had rained during the night. The hazy forms of some Thompson's gazelles grazed nearby on the wet stalks of the dead grass. I found Jason, Jewel and all four of their adult offspring around the den. Amba was sniffing around near the burrow entrance; Cinda was eating something in a patch of high grass. The others were lying resting. I watched as Amba put her head into the entrance of the den, and heard the soft whining as she called the cubs. She withdrew her head and stood, looking down, but no cubs came tumbling to the surface. Amba walked over to a nearby hole in the ground and again I heard her call. Just then Jewel walked up to the den and whined, but there was no response.

Suddenly Amba gave a few small whimpering calls and then, lifting her nose to the sky, she howled again and again. The other jackals looked up and, one by one, joined in. As I listened I suddenly knew that I should never see the tiny cubs again. To me, the mournful howls were the last call.

When all was silent Cinda reached down in the grass and picked up the thing she had been eating; it was the small dead body of a jackal cub. She carried it a short distance away and buried it. Was she a murderer within the family? Had she managed to kill her small siblings, one by one? It is possible, for most carnivores will occasionally kill and eat one of their own species. It is far more common, however, for such cannibalism to occur *after* an animal has died. Jackals are susceptible to various diseases: it seems more probable that the cubs died from natural causes. I remembered again how the two cubs had seemed so listless the day before, and how all four had so frequently stumbled over grass stems during their move to the new den – an indication, perhaps, of weakness.

Another factor which favours Cinda's innocence is that it is common for the older offspring of a pair of jackals, particularly the female offspring, to remain with their parents and help look after the next litter. These 'nursemaids' will help to chase hyenas and other dangers from the den, and they regurgitate to their small siblings quite as frequently as the parents (though often they join their siblings in begging from a parent, usually managing to get a share of the food). Over and above these duties, elder siblings usually play vigorously with the small cubs and spend long periods grooming them.

After the death of the five new cubs, Jason and his family moved back to the area of the fifth burrow and after a while I realised that this was their rendezvous for, although they usually hunted separately, they normally lay up in this spot. Wolves also have rendezvous areas where they meet each other after their cubs have grown up and left the den.

When I next returned to the crater, six months later, I went straight to the rendezvous, hoping against hope that I might find at least a few of Jason's family still together. Almost immediately I spotted one jackal curled up near some reeds beside a pool of water. I drove closer, checked up on the whisker pattern, and found that it was Jewel.

For an hour I waited with her during which time she made no movement other than a slight twitching of her ears as the flies bothered her. Then I spotted another jackal approaching. As it reached a slight rise in the ground, about sixty yards from Jewel, it stopped and looked around. It was quite close and after a moment I saw that it was Jason. Obviously he could not see his vixen who was still sleeping, and suddenly he pointed his nose up and howled five times. Jewel at once jumped up and gave answering calls and then, from the reed patch, I heard the howling of other jackals. A moment later three adults appeared and, rushing up to Jason with wagging tails, greeted him in turn, first nosing and licking at his lips, then flopping on to their sides in front of him and spreading their hind legs, tails still wagging vigorously. Jason groomed each one for a few moments and then trotted towards Jewel.

It took me several minutes to sort out the three young jackals, but soon I was certain they were Cinda, Amba and Nugget. At sixteen months they were fully grown and, indeed, Nugget was slightly bigger than his mother. A little later, though, I saw that, despite his superior size, Nugget still showed respect for his mother. Twice as Jewel was heading back to her resting-place, Nugget flopped on to his side in front of her, presenting himself to be groomed. Jewel, who was not, it seemed, in a mood for grooming, walked around him: the third time he intercepted her she snapped at him, biting his muzzle as she had done when he was but a cub. Then Nugget moved away with his tail tucked submissively between his legs and left his mother in peace.

I had a week at the crater. Each day I found several members of the Jason family resting at the rendezvous area or hunting on their own in Jason's home range. But although on several other occasions I found the parents and three of their offspring resting together, I never saw Rufus with them. I wondered whether an accident had befallen him or whether, perhaps, he had left his family for good. The others, though, were still united. One day a strange pair of golden jackals walked past, quite close to the rendezvous. Jason saw them immediately and hurried over, his hair bristling and a snarl on his face. He attacked the male of the pair and during this fight four of his family ran from the reeds and helped Jason chase the intruders right out of the territory.

On the last evening of my stay I found only Cinda at the rendezvous. As the sun set she got up and trotted off, occasionally pausing to catch an insect, until she reached the southern boundary of her family's home range. There she lay down and, curling up, closed her eyes as if to sleep. A few minutes later she opened her eyes again and looked around. Suddenly she tensed, staring southwards, and following her gaze I saw a male jackal approaching her, his hair erect. Cinda did not get up but lay with her ears pressed back against her head, her nose pointed towards the stranger.

Slowly he approached and then began to walk in a circle around Cinda, gradually getting closer and closer. After a few moments his hair went down and he went right up to her and briefly sniffed her rump. Cinda did not move. Then the stranger walked away, but only for ten yards before he too lay down. I waited with them for the last half-hour of daylight. Every now and then the male lifted his head and stole a quick glance at Cinda: each time she was curled up and looked asleep. When he was not looking she in turn stole quick glances at him.

It was frustrating to have to leave the next day. I drove out of the crater through Jason's territory, but saw no sign of Cinda or the male, and it was another six months before Jane and I were able to return for a week. And then, search as I would throughout Jason's hunting grounds and the country immediately surrounding them, I only found one member of his family: Cinda. Each day I found her resting at the old rendezvous – and each day a stranger male rested with her. A stranger to me, that is. I could not tell whether the male was the prospective suitor I had seen on my last visit, but he was a fine jackal, much larger than Cinda, the runt. It was November and the male was courting Cinda even as, two years earlier, I had seen Jason courting her mother.

Where, though, was the rest of her family? Perhaps Cinda's fine young male had challenged Jason, fighting for the right to court Cinda in the place of her birth, driving the older male from his hunting grounds. The other cubs might have their own mates too, their own territories to mark out and defend. I have not lost hope.

When last I saw Cinda she was curled up a few feet from her mate, near the fifth den of her cubhood. Jane and I had seen her courtship consummated – she at least lived to propagate Jason's fighting blood in her offspring. The tropical dusk was giving place to darkness, and I was about to turn the car and drive back to camp. Suddenly, in the distance, I heard a jackal howling. It was joined by another and then another. When the trio quietened their neighbours took up the strange high-pitched call, and then I heard more jackals to the south, and two more to the west. Finally Cinda and her mate howled, sitting side by side. Their duet, to my ears at least, was the last.

How difficult for man, despite his efforts, to learn the secrets of the animals he studies. The howling of the jackals, back and forth across the plains, probably shouted the information I so badly needed. 'Here am I, Jaaason. And Jewel toooo,' might have been the message from the west. And perhaps Nugget and his mate had answered from the east. But I was a mere human and it would take me months of research to piece together the information which Cinda, in those few moments, had stored away in her golden head. I sighed a little as I turned the car and drove back towards camp.

Spotted Hyenas

the Chuckling Hunters

Spotted Hyenas
the Chuckling Hunters

Bloody Mary and Lady Astor, leading matriarchs of the Scratching Rocks Clan, began to run fast over the moonlit plain, their tails aggressively curled over their broad rumps. Behind them ran some eighteen other members of the clan. About sixty yards ahead two hyenas of the neighbouring Lakeside Clan were resting close to the boundary of their territory. It seems that they were fast asleep, for when they got up Bloody Mary and Lady Astor were only a few yards from them. One of the pair was lucky and escaped, running for its life, but the other was not quick enough. Bloody Mary and Lady Astor seized hold of it and a few moments later it was practically hidden from sight as more and more of its enemies rushed in to bite and rend at its body. The night was filled with the fearsome roars and low whooping calls and growls of the triumphant Scratching Rocks Clan and the horrible screams of their victim.

Suddenly, however, a group of ten hyenas of the Lakeside Clan materialised out of the night and came racing in tight formation towards the battle ground. This group was small, but it was within its territory and the hyenas, as they ran to defend their 'rights', were confident and aggressive. The unruly mob of Scratching Rocksters retreated hastily, leaving behind their badly wounded victim. For a short distance the Lakeside Clan pursued them, but once they had crossed the boundary into Scratching Rocks territory they stopped, uneasy on foreign soil.

Meanwhile the Scratching Rocksters, once they were well within their own territory, also stopped, and the two rival clans faced each other, both sides keeping tight formation. Each individual held its tail curled stiffly over its rump, and the low growling whooping calls sounded louder and louder in the night air. And all the time both clans were swelling in numbers as more and more members, attracted by the calls of battle, hurried to the scene.

Suddenly I saw the shadowy forms of Bloody Mary and Lady Astor rush forward, side by side, and a moment later the rest of the clan was behind its leaders. For a short while the Lakesiders held their ground, and there were loud roars and shrill giggling, chuckling sounds as hyenas briefly attacked and chased each other in the skirmish. And then the Lakeside Clan retreated, running back into its own territory. After chasing for a short distance the Scratching Rocksters, who had once more crossed their boundary, began to feel uneasy and they stopped. Again the two clans faced each other, the whooping calls filling the air until the Lakesiders, reaching a peak of frenzy, rushed forward to renew hostilities. Another brief skirmish and then the Scratching Rocks Clan once more retreated into its own territory.

And so it went on, each clan surging forward in turn behind its leaders and then suddenly breaking and rushing back from the aggressive charge of the other. Eventually there were between thirty and forty hyenas on each side, and the cacophony of their weird calls, the rustling and pounding of their heavy feet, the menace of their dark shapes were everywhere around us in the moonlight.

Twenty minutes from the start of the affair the skirmishing suddenly ended and members of both clans moved farther and farther into their own territories, some occasionally glancing back over their shoulders as though to make certain there were no further infringements of the boundary. Hugo and I had seen a number of territorial disputes before, between the different hyena clans of the Ngorongoro Crater, but never one to equal that which we had just watched for sheer, seemingly unprovoked hostility. For, whilst the two resting Lakesiders who had started the incident may have been guilty of trespass, at the very most they were only a few yards on the wrong side of their clan boundary. What a price one of them had paid for its indiscretion, for almost certainly it would die of its wounds!

When the hyenas had dispersed Hugo and I left, for our son, two-year-old Grublin, was waiting for us back in the Munge cabin across the plains of the crater. On our way home we passed several Scratching Rocks hyenas and saw that a number of them limped and two had torn and bleeding ears. Our way lay across Scratching Rocks Hill, the small rise on the flat plain which has given its name to the clan of hyenas which we were studying. From the crest of the hill we could see the firefly twinkle of the Munge cabin light. Home.

At the time Hugo and I, with Grublin, and Moro and Thomas, our two African servants, were camped in the crater for the fourth of our hyena study periods. Hugo was spending much of the day analysing his jackal and wild dog material, and so was able to be in camp with Grublin. This left me free to study hyenas. Only in the evening, when we could leave Grublin with Moro safely inside the cabin, did we sometimes go out together. Then we had little to fear for our son's safety for the windows of the cabin would be barred, the curtains drawn, the strong door bolted.

Hugo and I had our supper in the cabin, with Grublin perched between us and trying to tell us, in the disjointed words and phrases of a two-year-old, all that had happened since we left. And soon we were in bed, with the leaves of the giant fig tree rustling overhead, the gurgling of the Munge River behind, and the moonlight paling the small squares of the cabin windows.

Each time we return to the crater it takes me three or four days to renew my familiarity with the hyenas of the Scratching Rocks Clan. Of the sixty or so clan members I know more than half very well, and I can identify most of the others by checking with photographs in my hyena recognition book. In this album are shots showing every hyena from both sides. It is not difficult, with practice, to identify the different individuals by the pattern of their spots, for like fingerprints in humans, each pattern is different. After a while many hyenas become equally familiar because of other individual characteristics such as the way they walk or hold their heads, or the shape of their bodies.

When I first started watching hyenas I was not, I must confess, particularly fond of them – although, from the start, I was fascinated by their social behaviour. But at that time I could sympathise with those people, which means most people, who have no soft spot for hyenas. That, however, was before I knew Bloody Mary and Lady Astor, old Mrs Brown and Nelson and the other members of the Scratching Rocks Clan. Every hyena has an individuality which belongs most peculiarly to itself and, after watching them for a while, I found myself becoming very fond of hyenas indeed.

Bloody Mary is the top-ranking female of the Scratching Rocks Clan, and since hyenas have a matriarchal society, Bloody Mary is the clan's leading hyena. She is blind in her left eye yet she commands the respect of all her subordinates, and when she charges in at a kill with her short tail erect and her mane bristling there are none who will gainsay her.

No mention of Bloody Mary is complete without a mention of Lady Astor, for most of the time the two do everything together. Lady Astor enjoys a position all but equal to that of Bloody Mary, and they look so much alike that I suspect, in fact, that they may be sisters. Both are aggressive, but Bloody Mary, perhaps because of her slightly superior social status, is calmer than her friend, less likely to squabble over trifles. Each of these females has acquired a girth as imposing as her social stature, and each of them undoubtedly weighs over one hundred and thirty pounds. When the two set off at the head of a group of their clan, intent perhaps on scent-marking one of their boundaries, they make an unforgettable picture as they walk or canter with their fat sides touching and their short tails tightly curled over their buxom jostling round rumps.

Since I have known her, Bloody Mary has borne twins, Vodka and Cocktail, and Lady Astor has had one daughter, Miss Hyena. But neither mother has permitted the business of child-raising to interfere with her social interests.

Mrs Brown, on the other hand, has been preoccupied with her cubs since I first got to know her four years ago. At that time she had just lost the end of her nose in some battle; the part that remained was conspicuous for it was bright scarlet. To-day the wound has healed, leaving her two nostrils bridged by space. Mrs Brown, whose social status is neither high nor low, is placid and elderly. Three years ago she weaned one cub, and promptly gave birth to another, Brindle. Whilst Brindle was small, his mother seemed content to spend hours lying near his den, suckling him frequently, passively watching his vigorous romping. To-day Brindle is twenty months old, yet Mrs Brown still nurses him and spends most of her time lying close to her overgrown baby.

Like Bloody Mary, Mrs Brown has a special friend, the ponderous old female, Baggage. Baggage, like Mrs Brown, is a motherly type and spends hours with her two cubs, Sauce and Pickle. Baggage is characterised by a pendulous belly which brushes the grass-tops when she wanders home from a kill, and by enormous liquid brown eyes. In addition, she has a passion for spring cleaning. She cannot resist digging out and tidying the entrance to the twins' den, even if it is just with a few scrapes of one paw, every time she arrives. I always picture Baggage permanently enveloped in a cloud of dust.

Her cubs, like most twins, are inseparable, but it is not difficult to distinguish between them, whether in looks or in character. Sauce is the bolder of the two, the first to run and greet a hyena visiting the den, the first to investigate any animal or bird that walks past. And everywhere that Sauce goes, Pickle follows, and everything that Sauce does, Pickle does too. And so the twins make a formidable pair, for if there is a squabble amongst the cubs they are united. Even fully grown hyenas often run off, their tails tucked nervously between their legs, when the twins charge together, their hair bristling with indignation.

The old male Nelson is blind in his right eye and his ears are tattered from endless bickerings over food and females, for he does not enjoy a very high social rank. He walks with his neck held rather stiffly and his head twisted slightly so that the good eye points ahead; this gives him a somewhat clownish appearance. Often, too, he trips over tufts of grass or other irregularities in the ground, and once, as he walked past a den, he stumbled right at the edge and vanished down the deep burrow head first.

But Nelson's most distinctive characteristic is his voice. It is difficult to describe, to those who have not heard it, the weird whooping 'ooooo-whup' of the hyena. Each 'ooooo-whup' forms part of a series of ten or more calls; they start loudly and often end very, very softly with a low, single-syllable 'ooooo'. It is the most frequently heard of the hyena's fantastic range of sounds and serves, amongst other things, as a means of contact between scattered clan members. Undoubtedly hyenas can recognise each other from their voices alone, for even I can identify many individuals in this way. Nelson has a rather pure baritone which he himself seems to enjoy: his calling can be heard on almost any social occasion, and quite often he walks along 'ooooo-whupping' quietly and seemingly to himself.

Until recently very little was known about the ways of the hyena. Most people pictured a cowardly scavenger existing off the remains of lion kills or other carrion, with perhaps the odd leather boot (stolen whilst its owner slept) thrown in. Some early naturalists wrote that hyenas sometimes hunted at night, in packs, but this aspect of their behaviour was overlooked by the majority of people until, recently, it was publicised by Dr Hans Kruuk, a young Dutch scientist working at the Serengeti Research Institute.

It was Hans's careful observations which first revealed that hyenas live in social units, which he named clans, comprising anything up to one hundred individuals. He found that the Ngorongoro crater was divided into eight different territories, and that each clan regularly patrolled its territorial boundaries, marking them at intervals with scent from glands situated under the hyena's tail.

From the top of Scratching Rocks Hill it is possible to look over the whole territorial hunting grounds – about six square miles – of the Scratching Rocks Clan hyenas. Ahead, if one faces the Crater Lake to the south-west, their territory stretches for over a mile across the flat plains, and to the left of the hill, too, the clan owns a small stretch of plains. Behind the hill the hyenas have access to a stretch of the narrow Munge River, and beyond that a fairly small area of the rolling hill country below the crater rim. About four hundred yards to the right of Scratching Rocks Hill the Munge River flows into the Munge Swamp, prior to emptying into the Crater Lake. This swamp, bristling with tall reeds, haunted by rhinos and reedbucks during the day, elephants and buffaloes at night, is much loved by the Scratching Rocksters for it is a cool place during the heat of the day. And here they are fairly safe from prying human eyes, for most of the reed beds are impenetrable by car. On the far side of the marsh a flat-topped hill, which we call 'Table Mountain', rises steeply, and the Scratching Rocks Clan territory extends right over this hill and includes a large section of the hill country beyond.

Three other clans have territories adjacent to that of the Scratching Rocks hyenas: the Lakeside Clan, the powerful Munge River Clan, and the Table Mountain Clan. The boundaries between the clans are not stable over long periods of time. Only a year ago, for instance, Table Mountain lay, as might be expected, within the territory of the Table Mountain Clan. But gradually, over many months, the Scratching Rocks hyenas managed to push their neighbours farther and farther back into the hills towards the crater rim. And this took place at a time when the Scratching Rocksters themselves were losing ground on their opposite flank, for the Munge River Clan was pushing its boundary across the plains towards Scratching Rocks Hill. Indeed, for a while Munge River marking parties were actually leaving their scent on the very top of the hill. But the Scratching Rocks Clan, having gained a huge piece of land from the Table Mountain Clan, then proceeded to push the Munge River Clan back to its old boundary. During one year, therefore, the Scratching Rocksters almost doubled their territory – I only wish I could have been there the whole time for there must have been many exciting battles.

The mothers of the Scratching Rocks Clan all raise their cubs within a relatively small area of the plains near Scratching Rocks Hill. The closest I have ever come to seeing a hyena birth was when, one afternoon, I met Bloody Mary carrying a new-born cub in her mouth. It cannot have been more than half an hour old, for it was still wet and the afterbirth was attached by the umbilical cord. I followed the mother and saw her carry the minute black baby into a large burrow in a disused termite mound. I waited there for the rest of the day, but when darkness fell Bloody Mary had not emerged from her den.

For the next two days Bloody Mary lay just outside the burrow nearly all day long; when she moved it was to go down into the den, presumably to suckle her cubs. Hyenas, unlike most carnivores, are born in an advanced state of development. Their eyes are open, many of their teeth are fully erupted, and they are surprisingly active, pulling themselves along with their front paws. Nevertheless, although I visited Bloody Mary's den for a while each day, I saw no sign of her cubs, until, when I drove up on the tenth day, I found them both suckling from their mother at the den entrance. They looked minute and very black beside the huge bulk of Bloody Mary, and they lay, one on top of the other, kneading her milk-swollen udder with their tiny paws.

When their feed was over, Vodka and Cocktail clambered over Bloody Mary's legs and she licked them briefly, one after the other. Their eyes were still misty grey-blue, and when they walked they wobbled and frequently stumbled, falling on their noses. Their tails were ridiculous short pointed spikes, and they had big feet and stumpy noses like the puppies of domestic dogs. Presently they wandered back to the den. Cocktail, whom I recognised from a tiny defect in one ear, went in carefully, but Vodka lost his footing at the edge and disappeared abruptly, sending up a cloud of dust.

For another week Bloody Mary kept her cubs on their own, and each day they became firmer on their legs, their eyes grew brighter, and they stayed outside for longer at a time. They became more playful, pulling at each other's ears and Bloody Mary's tail, and they showed increasing curiosity. One day an Egyptian vulture landed close to the den, searching for scraps. As it waddled along I saw first Vodka and then Cocktail poke a cautious head out of the burrow, staring with large eyes at the white bird. Very slowly, one after the other, they inched their way out and took a couple of inquiring steps towards their visitor. But when the bird accidentally moved towards them they tumbled over each other in a wild scramble back into the den.

During that week Bloody Mary's friend, Lady Astor, frequently kept the new mother company, lying close to the den. Often, too, Lady Astor's daughter, Miss Hyena, lay with the two dominant females. Miss Hyena got her name when I caught her staring at her reflection in a pool of water. She was about a year old at the time, and exceptionally pretty, with a pale creamy background to the bold black and glossy chestnut of her large spots. Her eyes sparkled and had a special lustre of their own, and her ears were neat and daintily set above her well-shaped head. She would, I decided, win a hyena beauty contest without any difficulty.

Bloody Mary, Lady Astor and Miss Hyena were the only guardians of the two tiny black cubs when, one moonlit night, a raiding party of seven Lakeside Clan males arrived at the den. Bloody Mary and her friends were resting as the group appeared, but they started up when the seven were still about twenty yards away. All three stared for a moment, and then fled. The Lakesiders continued on right up to the den, their tails curled. I had no idea what they were planning to do, but through my mind flashed a picture of a tame hyena cub, belonging to Hans Kruuk, which he had rescued from a marauding male hyena one night. That youngster suffered a gashed throat, a punctured windpipe and a broken jaw, and only careful nursing had saved its life.

Bloody Mary and Lady Astor did not run far before they stopped and turned to look back towards the Lakesiders. Suddenly Bloody Mary began to return. Either her maternal or her territorial instincts had reasserted themselves and her tail bristled with aggression as she ran towards the seven males, Lady Astor close beside her. Miss Hyena was too young to take part in such encounters. It was two against seven. I drove a little closer to the den. And now I shall never know whether it was the sudden noise of the engine or the approach of the two dominant females of a rival clan which sent the seven Lakesiders running back towards their own territory.

Bloody Mary and Lady Astor rushed after the males for a short way, growling loudly. Then they turned and ran back to the den, sniffing the ground, still growling. For a while they milled about there and then first Bloody Mary, followed by Lady Astor, started a long series of 'ooooo-whups'. Miss Hyena, who had returned, joined in. Next the two females ran over to a group of their own clan at a den some hundred yards away. They were still growling and it looked as though they tried to incite the others to accompany them on a mission of vengeance, for twice they started running in the direction of Lakesider territory, looking back over their shoulders at the other hyenas. But no one followed, and eventually they returned to lie close to Bloody Mary's den, where for another half-hour they continued to stare towards Lakesider territory and to utter low growls. It was probably no coincidence that, the following day, Bloody Mary moved her three-week-old cubs to share a den with Mrs Brown's Brindle and Baggage's two cubs.

At this point it might be of use if I were to interpose a word about hyena sex. It is not possible to distinguish, visually, between a male and a female hyena cub since the female is endowed with reproductive organs which appear, externally, amazingly similar to those of the male. The curious phenomenon has led to the fallacy that the hyena is a hermaphrodite. One trapper, for instance, was asked to catch six hyenas, three of each sex. He rapidly trapped three 'males' but had difficulty in finding even one female. Whilst he was still searching, one of his 'males' produced triplets.

The similarity between the sexes is enhanced by the fact that the openings of the vaginal and urinary tracts of the female are actually situated within the enormously elongated clitoris – which can be erected just like a penis. It is not until a young female has given birth that her two black nipples become obvious, and make it relatively easy to discern her true sex. The old females present no problem, for the prolonged suckling of a series of cubs ensures that their nipples become permanently stretched and extremely conspicuous.

Theoretically, the scrotum of a young male hyena should appear larger than its false counterpart on a young female. But it seems that my scrotum judgments are poor: recently a number of hyenas, such as Dracula and Hubert, have, by virtue of unexpected motherhood, become Countess Dracula and Huberta. And so, to make things simple, I have given the cubs neutral names, and refer to all of them, for the present, as 'he'.

The unsexable cubs of Mrs Brown and Baggage, at the time when they were joined by Bloody Mary's youngsters, shared a den right at the base of Scratching Rocks Hill. I called it the Den of the Golden Grass because, every evening, the setting sun burnished each blade of grass as if with gold. Mrs Brown's Brindle was about two and a half months old and beginning to lose his black natal coat: his head was a pale grey colour, as were his neck and shoulders where the first spots had made their appearance. Sauce and Pickle were a couple of months older, and only their rumps and legs were still dark in colour.

It was one evening, when Brindle was suckling from Mrs Brown and the twins were playing near the den, that Bloody Mary arrived with Vodka. The cub's neck was gripped between his mother's strong teeth, his body dangled. One of Bloody Mary's canines looked dangerously near his left eye, which was screwed tightly shut whilst he peered at the world, somewhat lopsidedly, through the other.

As Bloody Mary got close the twins darted down into the burrow, and Brindle, startled by their sudden movement, rushed after them. Mrs Brown remained lying on the golden grass. Bloody Mary walked right up to the den and dumped Vodka at the edge. He had no chance of getting his balance, but tumbled straight down and thus, somewhat unceremoniously, made his first acquaintance with his future den-mates.

Before setting off to collect her other cub, Bloody Mary paused to greet Mrs Brown. Hyenas have evolved a number of greeting ceremonies of which sniffing and licking the face and the genital area are the most common. As Bloody Mary approached, Mrs Brown, whilst remaining lying, pulled back the corners of her lips in a submissive grin, reminiscent of a nervous human smile. Bloody Mary put her nose down to the lower-ranking female's groin. Mrs Brown, as hyena etiquette demands, raised one hind leg and, as Bloody Mary did the same, both females sniffed and licked around each other's nipples. Then, with a few idle wags of her tail, Bloody Mary walked back towards her old den.

Ten minutes later she returned, Cocktail dangling. She placed him in the den and, flicking her tail a couple of times, lay down nearby. Soon the older cubs emerged from the den, Sauce first, followed closely by Pickle and finally Brindle. They stood in a row looking at the dominant female. Sauce, as usual, was the first cub to greet Bloody Mary. He went over to her quite calmly and pushed his black nose under her thigh. She raised her leg a couple of inches, but ignored the leg which he raised politely above her head. Pickle followed suit and the twins sniffed under Bloody Mary's leg together. Pickle also lifted a leg, but as he was safely on the far side of Sauce, Bloody Mary could not even see the gesture.

As the twins wandered away, Brindle, with infinite caution, approached Bloody Mary. His short neck was craned forward, his tail tucked tightly between his legs. He moved in a series of quick jerks, his grey head nodding rapidly up and down. Between each step he seemed poised for instant flight. At last he got close enough to stab his blunt nose dartingly towards Bloody Mary's thigh, but as the large female slightly lifted one leg in response, so Brindle bounced away in a series of tiny jumps, only to circle right round her and repeat the whole performance. This time he did not dart away when Bloody Mary moved her leg and, to my amusement, when this very small cub lifted his black leg she did respond, sniffing long and intently in greeting.

Brindle then returned to his interrupted suckling, and the twins began to play again, pulling at each other's ears and cheek hairs and tails, chasing each other in circles around the den and around Mrs Brown and her cub, tugging at opposite ends of a dead stick. Occasionally the black faces of Vodka and Cocktail appeared at the entrance of the den as the tiny cubs stared with large concentrated eyes at the boisterous twins, only to bob down hastily if the game got too close.

Hyena dens with resident cubs form the focal points in the social life of a clan. 'Visiting' usually begins at sundown and continues, on and off, throughout the night. Some hyenas just wander past, headed elsewhere, pausing only to greet their friends, but there are many who visit for long periods of time, lying near the den, playing with the cubs or adults already there, greeting new-comers. Usually a number of cubs, old enough to move around without their mothers, wander across every evening to a den where smaller cubs are living, stay there throughout the night, and return to their own home dens every morning. This is the typical, relaxed and happy-go-lucky pattern of hyena social life around the communal dens.

Lady Astor, Bloody Mary's friend, was the next hyena to arrive that evening. She greeted Bloody Mary and, when the latter gave the wide yawn so often shown by the dominant hyena of a greeting pair, Lady Astor, wrinkling her nose, licked right inside Bloody Mary's mouth. By now the cubs were clustered around Lady Astor, sniffing under her thigh: she ignored the raised leg which each one of them offered, but lifted her own, impartially, for all of them. Then, with a few tail flicks, she flopped down near Bloody Mary.

Miss Hyena wandered along after her mother, made the rounds of greeting, and then lay to suckle. Lady Astor raised one front paw as her daughter pressed close and laid it companionably across Miss Hyena's flanks.

Before darkness fell, several other hyenas had wandered up to the den. Five cubs of different ages arrived and joined the twins and Brindle in wild chasing games. Then old blind-eyed Nelson came, announcing his arrival with a series of 'ooooo-whups' which turned to high-pitched giggles as the cubs rushed to mob him. This treatment is often accorded to low-ranking males when they approach a den. Now, with Sauce and Pickle in the lead, and all with tails erect and bristling, the cubs raced towards Nelson. But, although he ran off for a few yards, he soon stopped and turned towards the youngsters, and their impressive display ended merely in a greeting ceremony. Some males continue to run, and then the cubs may actually bite them and chase them completely from the vicinity of their den.

Finally, as the light faded, the ponderous form of Baggage ambled up. The twins, wrestling and tumbling in the grass, did not see her coming: nor, it appeared, did she see the cubs for she went straight to the den, poked her head down and began to dig. She was still digging when the twins noticed her and rushed up to greet her. She licked them vigorously one after the other and then lay down, with a deep sigh, and began to suckle them. As usual, it took three or four minutes of wriggling and shifting before Sauce and Pickle had sorted out which of them should lie on top; as usual, Sauce ended up with the best position.

It is always frustrating when one has to drive away and leave the hyenas just as darkness falls, for it is at night that they are at their most alert and energetic. This is the time when the hyena is most likely to hunt large animals, to scent-mark his territorial boundaries, to mate his females. Luckily, however, there is the moon, and in the crater the moonlight always seems extra bright for it is reflected by the lake and by the sunburnt paleness of the grass and trapped within the grey and dark purple and black shadows of the crater's rim. In Africa the light of the moon is not a cold light: its silvery quality robs the world of its colours and substance but steals no warmth from the air. And if there is a feeling of enchantment that is surely because we humans are creatures of the sun – the hyenas, I am sure, find no mystery in moonlight.

It is only at night that I have watched adult hyenas playing vigorously. The first time that I saw such a game I shall never forget. I was parked at the Den of the Golden Grass (which was silver under the full moon), and Grublin, about a year old at that time, was asleep behind me on the big bed of our Volkswagen camping car. The hyenas were playing all around us. Lady Astor lay on her back with all four paws up in the air, whilst the twins jostled each other for a good grip on her right ear, two other cubs pulled, growling, at the hair on her neck, and Brindle hung on to her tail. Just as her daughter took a dive towards her, Lady Astor gave a violent heave and a shake and scrambled to her feet, the cubs scattering and sprawling on the ground. For a moment the big female stood there, her mouth slightly open in a foolish grin, and then, as Sauce and Pickle together pounced for her tail, she was off, racing in a wide circle over the short dried grass. The twins' jaws snapped shut in mid-air where Lady Astor's tail had been but a second before, and as they landed the cubs bumped into one another and sprawled on the ground. But in a moment they were up again and rushing after the other cubs who were already pursuing Lady Astor. It was a gay hunt in the silver moonlight.

This was not the only game in progress. To my right three adults ran round and round in ever-decreasing circles until, with a shrill, hysterical giggle from one of them, they tumbled into a heap of snapping teeth and sparring paws. As I watched, two other adults cannoned into the trio and joined the fray. Sudden little groaning grunts sounded from one of them and, briefly switching on my spotlight, I saw that it was Bloody Mary. She was being nibbled and chewed in her neck whilst, at the same time, her tail was being pulled. She sounded like someone who has been tickled beyond endurance as she lay, looking quite helpless, and giving little spasmodic jerks in the air with her paws. But a moment later she scrambled to her feet and then all five were off, galloping under the moon, and silent save for the thudding of their paws on the hard ground.

From the corner of my eye I saw a slight movement at the mouth of the den, and I needed no spotlight to tell me that this was one of Bloody Mary's small black cubs. The short stubby nose and round ears poked up from the safety of the underground refuge as the cub watched the wild activity around him and then, as Lady Astor raced by, he was gone again. A few minutes later one of them popped up again, and so it went on: occasionally both of the tiny cubs peeped out at the same time.

With startling abruptness the playing stopped. Each hyena stood motionless, staring in the direction of the lake. I strained my ears and heard nothing, but the hyenas must have heard the distant thundering of hoofs for suddenly Bloody Mary was off, running fast in the direction of the lake, closely followed by Lady Astor and the other adults. As I stared after them I saw, after a moment, the first brilliant flare of Hugo's powerful photographic flash. For I was not alone watching the hyenas. Whenever the moon was bright enough for him to drive after them, then Hugo was out too, following different individuals of the Scratching Rocks Clan, recording their behaviour, hoping that one of them, one night, would be the one to start a chase so that he could make observations and photographs from the very beginning of a hunt.

There were many occasions when I went with Hugo to watch the hyenas hunting. One of the most spectacular hunts took place when we were following Countess Dracula – whose upper lip is split into a terrible sneer. For several hours we drove slowly after her as she wandered along, stopping to greet others, occasionally joining up with another group, then going off on her own again. At last she lay down and seemed to go to sleep for half an hour before ambling on again. It all seemed quite aimless and we decided we had chosen badly when we picked Countess Dracula as our leader. But suddenly she started a hunt, racing towards a bull wildebeest. He ran off a short way, and as Countess Dracula raced after him, she was joined by three more hyenas. Suddenly the bull stopped and turned to face the hunters. It was a fatal mistake for, within a few moments, at least ten more hyenas had materialised out of the moonlight, and the bull was surrounded. He tried to make a break then, but as he moved four hyenas rushed towards his flanks, jumping up and biting at him. Again he stood at bay, wheeling to face them. Soon, however, he again made a break, and this time he managed to reach our Land-Rover. There he stopped and, pressing his rump against the car, he faced his tormentors: Hugo could have touched him through the open window. Minute after minute went by. The hyenas milled around growling and softly whooping, but they dared not start an attack: the bull stood, sweeping the air with his curved horns. But all at once, possibly because Hugo or I made some small sound, he turned and attacked the Land-Rover, his horns crashing into the door and denting the heavy metal. One horn went right into the window, missing Hugo's arm by inches. And so we were forced to drive off, and the wildebeest's fate was sealed for, as he ran from the moving car, the hyenas grabbed him. From the photograph Hugo took we saw, afterwards, that Lady Astor led the final attack. In a couple of minutes the wildebeest was down with the Scratching Rocks hyenas fighting over his flesh; a few minutes later he was dead.

Hyenas, like wild dogs, kill by disembowelling their prey, and it is just as horrible to watch. As we have said before in this book, it is unlikely that the prey suffers as much as one would imagine: nevertheless, I have found that the only way in which I can overcome my physical revulsion is to concentrate on the behaviour of the hyenas themselves. It is an incredible sight when the members of a hyena clan rush up, one after the other, and push in to get a share of the kill. That night the body of the bull was invisible, hidden by thirty or more hyenas, within seconds from the time he went down.

Bloody Mary and Lady Astor fed together, their fat sides touching and slowly expanding as they tore off and swallowed huge mouthfuls of meat. Miss Hyena, Lady Astor's daughter, ate close beside her mother. Mrs Brown, arriving a bit late, took a huge leap and, quite literally, dived head first into the middle of the scrum. Wellington, the clan's leading male, scrambled up on to the backs of his feeding companions and sat there for a moment until he spied a tiny opening through which he could squeeze. And, all round the close-packed mob, we could see late-comers wriggling towards the food on their bellies, pushing their way through the legs of their companions.

Soon the squabbling was fierce, and the sounds became louder and louder, the whoops and the growls, the sudden spine-chilling roars, the nervous chuckles and giggles. And, mingling with these calls, was the slurping, chewing, tearing, crunching sound of some thirty mouths feeding on flesh, skin and bone. Small wonder that, during such scrums, unintentional cannibalism occasionally takes place as ears and noses and paws get bitten, for after a while it must be difficult to differentiate between the flesh of the prey and the bloodsoaked, meaty hide of another hyena. That night Nelson acquired yet another notch out of one ear. I suspect that Mrs Brown lost the end of her nose during one of these hyena feasts.

Gradually the kill was torn apart and the sounds of the hyenas diminished. I saw Mrs Brown run off with a leg, hotly pursued by Lady Astor and Wellington. Nelson, giggling shrilly, his one good eye pointed forward, suddenly raced off with a piece of skin, Black Watch close behind him. Baggage made off with a shin bone, uttering low nervous chuckles and looking over her shoulder at Mrs Stink who was intent on taking the titbit away. Quiz, a young male, quietly tucked into a juicy piece of meat close beside our car, hidden from his clan mates. Miss Hyena darted off with the wildebeest's tail, but dropped it a few minutes later when she found there was nothing on it but hair. Low-ranking Old Gold sneaked off silently with a minute scrap of gut. It reminded me of a clearance sale in a department store, each customer frenziedly grabbing some bargain, sometimes dropping it after a second look. And, throughout, the darting quicksilver jackals nipped in and out between the hyenas' legs, seizing tiny pieces here and there, licking the blood from the grass. I counted eleven silverback and three golden jackals.

Fifteen minutes from the time when the bull fell only the head and backbone were left. Wellington and Honeydew tugged at one end and, all at once, the backbone became detached from the head. Bloody Mary, the dominant female, was left with the head. In the end, nearly always, Bloody Mary gets the head. It is her prerogative.

The satisfying part about a hyena kill is that nothing is wasted. Every scrap of anything that could possibly be eaten is eaten, and the grass is licked clean of the very last smear of blood. Only the wildebeest's mane, tail and beard, together with the horns and part of the skull, are left. And the horns, when finally abandoned by the adults, often get carried back to one of the dens where the cubs chew on them and play with them for hours.

Not all chases which the hyenas start are successful. Perhaps the most effective way for a hunted animal to escape is by running straight into the midst of a dense herd of its own kind, for sometimes this confuses the hunters. Occasionally a wildebeest tries to escape by leaping into the Munge River, but hyenas have no fear of water and will plunge in after their prey.

Once, when Hugo and I were driving along the Munge River,

we came across a group of hyenas milling about on the bank and staring down at the water. The bank, at that point, dropped down steeply to the water some four feet below. As Hugo and I looked, a sleek head, looking rather like that of a seal, broke the surface of the somewhat turbulent stream. The creature shook itself and, from the cloud of flying drops, the familiar face of Lady Astor gazed up at us. A second later another head surfaced, and Bloody Mary appeared. These two submerged almost at once, just as ponderous Baggage dived down from the bank, her splash wetting the hyenas standing on dry ground and causing a wave to roll down the narrow river.

We soon realised that the water concealed the body of some animal – probably one which had drowned there when trying to escape from hunting hyenas. The water was about five feet deep just there and it seemed that many of the hyenas, gathered on the bank, did not dare jump in – presumably from a fear of their superiors rather than of the water, for all the higher ranking individuals jumped in without hesitation.

As we watched Bloody Mary we saw her making careful movements with her front paws under water before plunging her head under and leaving just a tiny hump of her buttocks protruding. She came up with a mouthful of meat. Again she moved her paws about, and again she found meat, and it seemed that she was feeling around for the position of the carcass. Certainly she could not have seen anything in that churned-up muddy water. Later we noticed that those hyenas which did not make pawing movements first, often surfaced empty-mouthed.

All this time the swift-flowing stream was gradually moving the carcass downstream to shallower waters, and eventually it drifted to a place where the bank was less steep. Bloody Mary managed to get hold of a hind leg and bring it above water and then, inch by inch, the hyenas dragged the dead body of a wildebeest on to dry land. Immediately the typical growling, giggling scrum formed around the kill. But there was one difference for, as lower-ranking individuals fled with tit-bits, looking nervously over their shoulders, again and again they fell head over heels down the bank and into the water. Sometimes they were dunked a second time as their pursuers fell into the stream on top of them.

In the crater the hyenas of the Scratching Rocks Clan hunt wildebeests frequently, and zebras rarely, but the neighbouring Munge River Clan commonly hunt zebras in the hilly country behind the Munge cabin. Once I went with Hugo when he followed members of the Munge River Clan on such a hunt. It was a brilliantly clear night, but the hill country was treacherous with sudden holes hidden in the long grass, and we tightened our seat belts and settled our crash helmets more firmly into place. We were following a group of thirteen hyenas, for whilst one or two will start a hunt for wildebeests, it is usually a much larger hunting party that sets off after zebras. Suddenly the hyenas began to run, swerving and heading straight up the hillside. We followed, and I expected the Land-Rover to topple down the steep slope at any minute. However, we reached the top without mishap and found the most incredible picture spread before us.

The scattered groups of zebras had massed together into a huge striped band, well over two hundred strong. As this herd cantered along we could see several stallions defending the rear, pausing every so often to bite back at the leading hyenas, their ears pressed close to their necks. The air was constantly vibrating to the shrill braying barks of the hunted zebras, and their hoofs thundered across the moonlit hillside. It was impossible to make an accurate count of the hyenas that were, by then, bunched up and strung out behind that herd, but there must have been over thirty. I saw one of them suddenly fly up into the air, about a foot higher than the zebras' backs, and I guessed it had been kicked. It rolled over twice when it landed, but got up and went on running.

For about fifteen minutes we followed the hunt and then, in ones and twos and small groups, the hyenas dropped back, giving up in the face of that solid wall of zebras and the active defence of the stallions.

Zebras are definitely a tough proposition for the hunting hyena and, invariably, the hunters give up if a zebra they are chasing stops running. It seems that, like wild dogs and wolves, hyenas are prepared to risk kicks from large prey, but are afraid of teeth and horns if the quarry turns to face them. We saw one lame zebra stallion chased by hyenas for four nights running: each time, after a short run, the stallion stopped and turned to bite towards the hyenas. The hunters stopped too and, after milling around for a while, trailed away and left their intended supper grazing in peace.

The easiest time of year for the hyenas of the crater, so far as food is concerned, is during the calving season of the wildebeests. It is also the time that is the most damaging to the hyena's reputation. For it is almost impossible to watch the capture and killing of a wobbly-legged newly born calf without feeling intense sympathy for both the calf and its mother. Time and again Hugo and I observed the birth of a calf, watched the mother lick its wet fur, laughed at its first struggling attempts to gain its long legs, marvelled at the short space of time during which it learnt to use those legs and run – and then noticed the approach of one or more prowling hyenas. Less than ten minutes from the time of its birth it might be caught and eaten for, during this rich season, the hyenas hunt as much during the daytime as they do at night.

It made a deep impression on me when I was working at the crater a month before my own son was born: I could not help dwelling on the seeming waste of effort and life: the cow wildebeest carried her calf for so long, at last she delivered it and then, in a flash, the newly arrived miracle was dead before having a chance to live. An unscientific way of viewing nature, but pregnant women must be allowed their imaginings. And sympathy, within bounds, cannot be wrong. Even the most devoted cat lover usually spares a thought for the fledgling bird laid out so proudly on her sitting-room floor, but she will not condemn her cat. And her cat, unlike the much criticised hyena, does not even need to hunt for its food.

There is another point to consider. It is not entirely without danger to itself that a hyena captures a wildebeest calf, for the mother usually defends her youngster with determination and vigour. I watched one spectacular encounter between an enraged cow and the dominant male hyena of Scratching Rocks Clan, Wellington.

As Wellington ran towards the mother and calf they cantered off, but as he got closer the cow turned and charged, bowling him over, wheeling and running off again. In a trice Wellington was up and chasing. This time he caught up and grabbed the calf in the shoulder, but once more the mother spun round and, with a powerful toss of her head, she sent Wellington, together with his victim, flying some four feet up into the air. As Wellington hit the ground the mother was ready for him and, down on her knees, she butted him against the grass, holding him there with her curved horns. She fought a losing battle, however, for the calf was wounded, and when Wellington finally evaded the mother he had no difficulty in grabbing it once more and running off with it. He held it by the neck and killed it with shaking movements of his head and a crunching of his powerful jaws. This time the mother wildebeest merely galloped in a circle around the hyena: perhaps she failed to charge because her calf was no longer bleating. Anyhow, as three more hyenas raced to the scene, the mother just stood and watched as her youngster was torn to pieces. Then she turned and wandered away grazing, to all outward appearances grieving not at all. When Wellington finally moved off I saw that he walked with one front paw held off the ground. He would not hunt again for several days.

Calves do, quite often, manage to escape from hunting hyenas. Sometimes this is when a bull wildebeest intervenes and charges the hyena, thus giving the mother and calf a good start. It is not necessarily from altruistic motives that a male wildebeest behaves thus – a belligerent bull is liable to charge a hyena whether the latter is pursuing a calf or not. Once, indeed, Nelson was charged four times, and bowled over once, when he was trying to eat a young Thompson's gazelle which he had caught.

It is when a mother wildebeest gallops straight on, rather than stopping to attack a pursuing hyena, that her calf has the best chance of escaping. For usually a calf will turn when its mother turns, circle her whilst she attacks, and thus become an easy prey once the hyena has eluded the cow wildebeest. Once, however, a very young calf continued to run whilst its mother spun round and bowled over one of a pair of chasing hyenas. Even as she did so the calf passed through a herd of zebra, one of which, for some reason which must remain a mystery, appeared to kick out at the fleeing youngster. The calf fell sprawling to the ground and it seemed certain that the second hyena, which was now only twenty yards away, would make an easy kill. But, even as the calf scrambled to its feet, the mother charged the hyena and knocked it to the ground. For a moment she held it there with her horns whilst her calf continued its interrupted flight. The mother then galloped on and was soon running fast and smoothly close to her youngster. The first hyena had given up – perhaps the cow wildebeest hurt it when she charged. The second continued the chase for several minutes but then it seemed to give up, falling farther and farther back and finally stopping.

Wildebeest calves are not, of course, the only young animals which a hyena will attack: he will go for the young of almost anything, including lion cubs and even rhino calves. A mother rhino usually keeps her new-born hidden in thick vegetation until it is about two months old – by this time it is already too large for a hyena to tackle. But one mother rhino, in the crater, brazenly wandered on to the plains with a tiny month-old calf. She did not keep it long: early one morning we found some fifteen hyenas of the Scratching Rocks Clan surrounding the mother and baby. When we drove closer we saw that the baby had a broken leg, but whether or not the hyenas were responsible we shall never know. Gradually during the morning most of the hyenas dispersed, but a few, including Wellington, Nelson and Baggage, stayed with the wounded calf, rushing in every so often to bite at it. The mother, who had probably defended her offspring all through the night, became more and more lethargic and eventually Dr Goddard, who was studying the rhinos in the area, shot the baby and ended its sufferings.

Two years later, to our surprise, the same mother rhino appeared on the crater plains with another month-old calf. Each night, for at least three weeks, groups of Scratching Rocks hyenas harried the pair, following behind, standing around in a circle, making sudden runs at the calf. But this time the mother defended her offspring successfully. The calf itself, though no larger than a hyena, and with no vestige of a horn marking its baby nose, charged again and again at its tormentors, its tail stiffly erect, and backed by the ponderous bulk and formidable horn of its watchful parent. Eventually the calf outgrew the danger period, and the hyenas lost interest in it.

Ngorongoro Crater is famous for its rhino; they can be found on the open plains, in the reeds, and in the hills and forests of the crater rim. I never worried at all about rhinos until those moonlit nights when Grublin slept in the back of the Volkswagen whilst I watched at the Den of the Golden Grass. Most of the rhinos are placid – so many tourist cars drive up to and around them each day that they would soon succumb through exhaustion if they charged each one. But just at that time it so happened that a 'bad' rhino had moved down from the crater rim. Three times Hugo was charged by him; once he had only just got away in time.

One evening when the twilight had, almost imperceptibly, given place to moonlight, there was a sudden loud snort close to the car. I was parked by the hyena den, with the light on and the gas burner lit, giving Grublin his supper. Quickly I switched off the light. A large rhino was standing some six yards from the car. As I watched, he snorted again, tossed his head in the air and pawed at the ground with one front foot. Then he took a few stiff bouncing little steps sideways, with his tail straight up in the air. Probably he was trying to decide exactly what the immobile car was, and what he should do about it. Any sudden sound might provoke a charge, and he was much too close for me to risk switching on the engine and trying to drive away. Carefully I turned off the gas and somehow, without alarming him, I managed to persuade Grublin that we had to be quiet. Together we watched the rhino.

After eight minutes, which dragged past like eighty, our visitor, with a good deal of pawing and horn-tossing, decided to leave. As he went he paused frequently and turned to stare at the strange white shape of the car that gleamed in the moonlight. The hyena cubs who had all stayed very close to the den during the visit now grouped themselves behind the twins, and wandered for a short way after the rhino. Soon they stopped to savour the delight of the visiting card he had left – a heap of steaming dung. For hyena cubs like jackal cubs, consider the fresh droppings of vegetarians a delicacy.

It is a pity that Grublin will not be able to remember those nights he spent at the Den of the Golden Grass. He often fell asleep watching the moonrise or the stars, and if he woke it was to hear the lowing, honking grunts of the wildebeests, or the strange howling cry of the jackals or, perhaps, the 'ooooo-whup' of a passing hyena. Of course, I did all I could to prevent him from waking, for it disturbed my observations. So if he stirred I tried to lull him back to sleep with a song. Sometimes, when the hyenas were active at such moments, I simply sang my observations into my tape-recorder. The results, played back the next day, were often hilarious. Imagine, for instance, to the gentle melody of Brahms' 'Cradle Song', 'Bloody Mary arrives, her face and neck red with blood, a leg in her mouth' or, intoned to the tune of 'Rock-a-bye Baby', 'Nelson gives a hysterical giggle as the twins lick his penis'.

One thing which threatened to disturb Grublin's slumbers every night was the vigorous attempt made by every hyena cub, and many of the adults, to devour the Volkswagen. On and off throughout the night the hyenas renewed their attacks, chewing and biting on the lights, the wires, the exhaust – anything, in fact, on to which they could get their teeth. Once Pickle came from under the car making awful faces, pawing at his mouth and shaking his head. I discovered why when I drove home and found the brakes wouldn't work: Pickle had bitten through the brake cable and got a mouthful of brake fluid.

The twins were the worst offenders. Once, when Sauce kept chewing one fog light whilst Pickle gnawed at the other, I tried everything to stop them. They completely ignored my switching on the lights and the engine, and merely cast bleary eyes up at me when I leaned out of the window and tapped the car. I then squirted Pickle, who was my side, with aerosol flyspray: he backed away, sniffing the air carefully, but returned almost at once. Finally I had to open the door and put one foot out: first Pickle and then Sauce scuttled back to the den, where they got involved in some game which, temporarily, distracted them from the car.

However, it was not the twins or any of the other cubs who most frequently woke Grublin, but the older hyenas. When they clamped their powerful jaws around some protruding piece of metal, and wrenched, it often sounded as though the car was being torn apart. The worst offenders, because they spent so much time at the den, were Mrs Brown's elder son, Master Beige, and Miss Hyena, Lady Astor's daughter. These two visited the den for hours every night, and they were amongst the most playful of all the hyenas. Once they had a twenty-minute game, half chasing, half tug-of-war, with a wildebeest tail, and once they played for a long time with a brightly coloured wooden truck which Grublin had dropped out of the window.

Master Beige also played frequently with his small sibling, Brindle. At that time Master Beige, though he would fill out a little, was full grown, whilst Brindle was still only about three months old. The game began when Brindle pounced on a rather large and very smooth roundish stone. Opening his mouth as wide as it would go, he struggled to pick it up, but it was just too big and his teeth kept slipping. Just as it seemed that he had, at last, managed to get a grip, Master Beige ambled up, seized hold of his small sibling's ear, and pulled. Brindle sprawled on the ground, but in a flash he was up and again trying to pick up his stone. But Master Beige, by worrying the scruff of Brindle's neck, yanking at his ears and pulling at his cheek hairs, made it impossible for Brindle to get a grip on his plaything.

Suddenly Brindle left the stone and darted to seize his tormentor by the tail, hanging on tight with his sturdy legs braced as Master Beige turned to bite at him. Gradually they moved farther back, Brindle still gripping his brother's tail. And then, as fast as his short legs would go, Brindle ran back to the unguarded stone. But Master Beige was too quick for him and grabbed the toy himself, running off in the moonlight with his head turned back over his shoulder, inviting pursuit. Brindle lumbered along behind. Master Beige slowed his gait until his sibling could bite at the stone in his mouth, and then he dropped it on to the ground. Brindle once more tried to get a grip on it, but always, if it seemed he might actually pick it up, Master Beige pounced, picked up the stone himself and loped off with it for a few yards. Brindle followed, Master Beige dropped the stone again, and so it went on.

After a while Brindle, it seemed, gave up. He wandered away until he reached a low thorny plant and pulled on a twig until it broke off. Master Beige watched him. Brindle shook his twig, dropped it and pounced on it: Master Beige could resist no longer. He loped across to join his sibling in a tug-of-war. But the moment one end of the twig was firmly in Master Beige's mouth, Brindle dropped his end and raced back to the stone. Still, of course, he could not pick it up, but this time, as his big brother bounced back to grab the toy, Brindle firmly sat on it.

Usually, when the cubs were playing, Mrs Brown was stretched out nearby. Old Baggage spent a good deal of time at the Den of the Golden Grass too, not only because of her cubs but also because she liked the company of Mrs Brown. Often, on a hot day, the two large elderly matrons wandered off together to cool themselves in the oozing mud of the reed beds. Still together, they would appear, wet and bedaubed with slime, at about four or five o'clock in the evening. Walking slowly up to the den, the two mothers would peer down, sometimes giving the short low-pitched call which brings the cubs scrambling to the surface. Baggage nearly always dug at the den entrance whilst waiting for the twins. Then the two females, followed by their respective cubs, would lie down in small depressions near the den to suckle their offspring – a meal which would continue for one to one and a half hours.

Whilst her cubs were under two months even Bloody Mary – who as number one female of the clan had many things to occupy herself with besides her children – spent a good deal of time near the den. This was probably because small cubs suckle more frequently, though for shorter periods of time, than cubs of three months or more. And so, whilst Brindle and the twins usually nursed only three times in twenty-four hours, Bloody Mary's, at that time, suckled twice as often.

Vodka and Cocktail, from the start, had extremely different personalities. Cocktail was, in all ways, quieter and less boisterous. He spent minutes on end grooming around Bloody Mary's ears and neck when she lay near the den, and would then sit in front of his mother, his chin in the air, inviting her to reciprocate. Bloody Mary usually responded and groomed him for a short while, nibbling at his fur with her incisor teeth or licking him all over with her rasping, cat-like tongue. Meanwhile Vodka would play, tumbling around with Brindle or clambering about on his mother and trying to persuade Cocktail to join in some pursuit more active than grooming.

As the two cubs grew older I noticed that they moved about together rather less frequently than most twins. This, I think, was because Vodka, with his bolder temperament, was eager, from the time when he was about four months old, to leave the den and follow Bloody Mary around. Even at that age, when the back half of his body and his legs were still black, he appeared from time to time at kills that were made not too far from his home den. Secure in the protection of the clan's top-ranking female, Vodka managed to seize juicy pieces of meat from under the noses of other adult hyenas. Often he fed whilst actually lying underneath Bloody Mary. Indeed, unless he was right beside his mother he could not safely eat his spoils. Once I saw him following Bloody Mary from a kill with a huge piece of skin in his mouth. He held his head high to keep his booty from trailing on the ground, but nevertheless he occasionally tripped over it and sprawled on his nose. After a few yards he paused to chew at the food and the group of five three-quarter grown hyenas which had been following him and casting longing glances at his large share, drew close. Bloody Mary continued and when she was six yards away one of the youngsters darted at the food. With a shrill giggle Vodka caught up the skin and, with his tail erect and bristling, hurried after his mother, twice tripping over his trophy and still followed by the hopeful

1. Bloody Mary, the dominant female of the Scratching Rocks Clan, leads her group during a border clash with a rival Clan.

2. Nelson.

3. Countess Dracula.

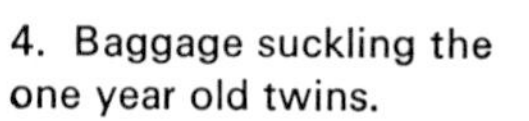

4. Baggage suckling the one year old twins.

5. Mrs. Stink.

6. Lady Astor.

7. Miss Hyena, daughter of Lady Astor.

8. A young hyena kisses an adult female in greeting.

9. Cubs greeting adults 'visiting' their den in the evening.

10. Bloody Mary and her cubs, Vodka and Cocktail, at the Den of the Golden Grass.

11. The cubs pursue Nelson. In the background Baggage sits up but is too full to join in.

12. Master Beige playing with his small brother, Brindle.

13. Sauce chases Vodka.

14. Mother hyenas frequently move young cubs from one den to another.

5. A one month old
ub stays close by its
nother.

6. Miss Hyena pushes
Master Beige as he
layfully pulls her ear.

17, 18, 19. Hyenas
harass a month old rhino
calf despite the
proximity of its
formidable mother.

20, 21, 22. Hyenas of the Scratching Rocks Clan pull down a wildebeest. Five minutes later a group of Lakesiders challenges them.

23, 24. Hyenas flee as a lion bounds over the Scratching Rocks. Old Gold trapped against the rocks, is mauled and choked to death.

25, 26. Wellington captures a wildebeest calf despite a determined defence by its mother.

27. Brindle, at four months, demonstrates the well developed muscles of a hyena's neck and jaw.

28. Eighteen month old cub in a weaning tantrum.

29. Hyenas dragging kill out of water.

30. Bloody Mary (in foreground) and Lady Astor chase Mrs. Stink from a kill.

31. Wellington keeping cool on a hot day.

32. Black Watch scoops earth over himself with front paw as a protection from flies.

33. Young hyena leaves kill with his trophy, a wildebeest's tail.

youngsters. But when Bloody Mary lay down, Vodka pressed close against her to eat, and the older cubs gave up and wandered away.

When Vodka and Cocktail were ten months old, and half grown, they began to move around together more frequently. Possibly the things that Vodka wanted to do, such as joining in at kills or going off on small expeditions of his own away from the den, no longer seemed so alarming to Cocktail as he grew older. Also, Vodka had calmed down after the exuberance of his extreme youth, and was willing to spend ten minutes or so at a time in grooming and licking with his sibling.

Nevertheless, I had no inkling that anything was amiss when I arrived one morning at the den which the two cubs had been sharing and found Vodka on his own, with Bloody Mary lying nearby. Vodka was restless. He kept walking around, sniffing the ground, or standing and staring across the plains. Suddenly he began to run fast, away from the den and his mother. Bloody Mary raised her head to stare after him, and when he disappeared over a slight rise in the ground she got up and wandered after him. I followed her, and presently we caught up with Vodka, who had arrived at a den which he and Cocktail had shared a few weeks earlier. Bloody Mary flopped down, but Vodka was still restless, and after ten minutes he again set off, racing over the plains. I followed, and saw him arrive at yet another den. He vanished inside it and presently a spray of dirt flew from the entrance as though he were clearing out the den. Bloody Mary again wandered up and lay down but, as before, Vodka was soon running off to yet another den. By this time I was convinced that he was searching for his sibling. Again he dug out the den – it was as though he were making quite certain that Cocktail was not hiding down there.

It was two hours before the cub gave up and lay down to rest beside his mother. During those two hours he ran from den to den, digging at them, sniffing into the air or along the ground. Many of his journeys were hazardous in the extreme, for it was the rutting season of the wildebeests and, in the daytime, the bulls charged any hyena that passed too close. Vodka, looking very small, just ran in huge semi-circles around the aggressive wildebeests, but Bloody

Mary, when following her cub, was frequently forced to turn and hurry back to the safety of the den she had just left. Once she was routed six times by the repeated charges of the same vigorous bull before she finally managed to catch up with Vodka.

Looking back over the events connected with Cocktail's disappearance, it seems that Vodka, that morning, *knew* that something had happened to his sibling. Bloody Mary, although she followed her cub from den to den, never joined him in his frenzied searching and appeared quite calm. But that evening, Bloody Mary too must have realised that something was wrong, for then it was she, with Vodka following, who went from den to den, putting her head down each one and calling softly. Occasionally, too, as she walked along, she uttered long series of mournful-sounding 'ooooo-whups'. For the next two days Bloody Mary continued searching, though rather half-heartedly, in different dens, but after that it seemed that she gave up. I never saw Cocktail again.

It is not at all unusual, in the crater, to see a hyena – a mother, a sibling or a friend – move from den to den as it searches for a particular cub. This is to some extent because, from the age of about four months, the cubs may move from one den to another on their own initiative. Often they stay only for three or four days before moving on again, although sometimes a den may be inhabited by the same cubs for three or four weeks (as was the Den of the Golden Grass). Younger cubs are frequently carried or led from place to place by their mothers. Often, I feel sure, a hyena moves from a den because an individual arrives there with whom it 'does not get on', or in order to be with a friend in a different den. Imagine, for instance, a den, or cluster of dens, housing a number of cubs of different ages. A quarter-grown cub, who has wandered over to play for several nights running, decides to stay rather than return to his original den. This means that his mother will probably lie at the new den. But another mother, who was there before, is subordinate to the new arrival and uneasy in her presence. So she carries her two small cubs, one by one, to another den some distance away. Small black cubs are attractive to older ones, and soon, therefore, older cubs arrive at this den and take up residence with the youngsters. Or another mother, who enjoys the company of the mother of the two black cubs, shepherds her own youngsters to that den. And so it goes on, round and round the available dens, year in, year out.

On one occasion I was responsible for a den move which, for the mother, must have been a most harrowing experience. Early one morning I drove up to a den and found the young Countess Dracula lying there, suckling her twins. They were her first cubs, only three weeks old, and she was still extremely wary of my car. As I drew up she stood and looked at them. Her face, with its grotesquely split upper lip, looked somewhat sinister in the grey light of dawn. Presently she bent to pick up one cub by the scruff of its neck and set off with him, occasionally putting him on the ground in order to get a better grip. She reached her destination – a den some hundred yards distant – and disappeared under the ground, still with Cub One in her mouth. Soon she reappeared without him.

Meanwhile, near me, Cub Two was uttering shrill whooping howls from within the den. Countess Dracula emerged from the far den and began hurrying back towards him. She ran the last thirty yards and the howling stopped as she stuck her head into the den entrance. But at the far den Cub One began to howl, loud frantic-sounding calls of distress. Countess Dracula jerked her head out of the den, stared towards the sounds, and began to run back to the yelling cub. She vanished under the ground.

The moment she left, of course, Cub Two started to call again. For a short while the harassed mother remained inside the far den but soon her head appeared and she stared back towards the crying cub. A moment later, as she started to return to Cub Two, Cub One clambered from the entrance of the far den and began to follow his mother. Countess Dracula stopped, looked at him for a moment, and then continued on her way. The tiny cub stumbled after her, frequently tripping and falling on his nose.

All at once a series of melodious 'ooooo-whups' drew my attention to the arrival of Nelson. He wandered towards the far den, peering curiously with his one eye at the tiny black cub: the cub instantly tumbled back to the safety of its burrow. Countess Dracula looked round, saw Nelson and simply raced towards him. With a loud growl she flew at him and chased him, giggling, away. Mothers of small cubs invariably keep adult males at a respectable distance: there are rumours of cannibalistic fathers.

Now, as Countess Dracula stood, looking rather helpless, at the far den, Cub Two, near me, who had been quiet for a while, started howling again. His mother at once started hurrying towards him. Again Cub One emerged from the far den and started to follow. This time, when he was about ten yards from his burrow, Countess Dracula turned, shepherded him back with her nose, and pushed him down. When she looked up it was to see Nelson approaching her other den. She raced back over the hundred yards and again chased Nelson away. Then she put her head down the near den. It took her two minutes to get hold of the cub who was squealing frantically down in the den. When his mother finally raised her head with Cub Two in her jaws, I could see that he was about to slip and fall. For a moment she put him on the ground to readjust her grip: instantly Cub Two darted back down the den. Countess Dracula glanced to the far den where her other cub was sitting and making the most appalling racket, took a step in that direction, glanced at Nelson who was standing not far away, changed her mind and dived right down into the near den. A moment later she emerged with Cub Two firmly in her mouth and ran the whole distance to the far den. The indignant squeals of the infant as he swayed to and fro in her jaws only stopped when she dumped him into the burrow with his sibling.

It had taken Countess Dracula no less than thirty minutes to move two small cubs one hundred yards. She lay down and, at last, the cubs were able to resume their interrupted feed.

There are other occasions when we would expect the mother hyenas to move their cubs and are surprised to find that they do no such thing. There was, for instance, the time when six lions came wandering across the plains towards the Den of the Golden Grass. It was in the evening and Bloody Mary, Mrs Brown and Baggage were all lying near the den suckling their cubs. Bloody

Mary sensed the approaching lions first and leapt to her feet, sounding the soft rattling alarm growl. Instantly all five cubs raced to the safety of the den and, after staring at the lions for a moment, the three mothers ran off for some fifty yards and then stopped. The lions went right up to the den and sniffed intently down the burrow. The five lionesses of the pride soon moved on, but the lion remained at the den for five minutes, and even made a few idle digging movements with one paw before he finally left.

During this time the mother hyenas just stood watching: not until the lions were all about two hundred yards away did they rush back to their cubs. After poking their heads down into the den, presumably to reassure themselves that their youngsters were all right, they spent five minutes running hither and thither, sniffing the ground and scent-marking in umpteen different places. After a few minutes the cubs ventured out again, and they too joined their mothers in marking the area with scent. It looked as though the hyenas were trying to get rid of the terrible smell of the lions by covering it with their own. Yet, to my surprise, no one moved from the Den of the Golden Grass.

It was far worse for all the hyenas concerned when a pair of courting lions stayed, for two days and three nights, above a cluster of dens that sheltered thirteen hyena cubs who, during that time, were imprisoned below ground. Their mothers, who wandered up from time to time and hung around in the background, were unable to suckle their offspring. Courting lions seldom feed during the ten days or so when the lioness is in heat – they just lie around and, at frequent intervals, mate. During the time this particular pair lay above the hyena dens it was raining and cool: there was no fierce sun to drive them, during the day, into the shade. Probably they were quite unconscious of the fact that they were disturbing the life of so many of the Scratching Rocks hyenas. Not that it would have bothered them had they known.

The relationship between hyenas and lions, especially in the crater where hyenas are so numerous, is fascinating. The hyena has been so frequently depicted as a scavenger at the lion's table that it may still come as a surprise to many, despite the recent revelations of Hans Kruuk, that the lion very often scavenges from the table of the hyena. It was three years ago that Hugo and I first saw this for ourselves. A group of Scratching Rocksters made a kill at sunset and when we arrived at the scene there were still only seven or eight hyenas around the dead wildebeest – for it is only at night, when all the hyenas are awake and alert, that the clan members congregate so rapidly around a new kill. A few moments after we had stopped the car a young male lion bounded towards the kill. He was about two years old, just beginning to sprout his mane. As he leapt forward, with a ferocious growl, the hyenas rushed back, but reinforcements were coming, and in a few moments a group of fourteen hyenas, uttering low whooping calls, began to move back towards the carcass, their tails erect and bristling.

Suddenly we noticed Lady Astor creeping up behind the feeding lion. She really was creeping, putting each foot down very carefully and taking, it seemed, great pains to preserve an element of surprise. She reached a point about five feet behind the lion, darted forward, bit the lion's foot and then rushed away, giggling shrilly as though overcome by her own daring.

The lion spun round with a roar and, as he did so, the other hyenas moved in and began to feed on the kill. The lion chased a short way after Lady Astor, and when he returned only a couple of hyenas ran from the carcass. Indeed, it seemed that the lion was afraid, for he made no attempt to drive the hyenas away again, but stood lashing his tail and rubbing his bitten foot on the ground. Even when a second two-year-old male lion arrived, the pair did not chase the hyenas away, but merely joined them at the carcass. It was a sight I shall never forget – the two lions feeding at the neck of the dead wildebeest whilst an ever-growing mob of yowling, giggling, growling hyenas fed opposite them.

However, although hyenas may sometimes lord it over young lions or single lionesses, they show a good deal of respect for a full grown male lion. One night Hugo and I, with Grublin asleep on the seat between us, followed Mrs Brown as she ran across the plains. Presently we saw the shapes of many hyenas ahead of us in the bright moonlight, and Mrs Brown joined them. We stopped close by. Fifty yards from the hyenas there was a pride of lions feeding on a zebra kill; two lionesses and a number of half-grown and quarter-grown cubs. A few jackals were close to the kill, but the lion cubs constantly ran at them so that they got few if any scraps.

After a few minutes Bloody Mary appeared from the shadows and stood near the car, her tail erect and bristling as she stared towards the lions. She began growling softly, and presently Lady Astor joined her. The two big females occasionally dug at the ground, first with one front paw and then the other, a sign of frustration. Then Bloody Mary gave a series of low 'ooooo-whups'; soon Lady Astor joined in, and then another hyena, and another, until the weird chorus was all around. A few moments later Bloody Mary and Lady Astor started forwards in the direction of the lions with the rest of the clan following closely behind. We counted twenty-eight Scratching Rocksters advancing on the pride, and as they got closer so they called more and more loudly, their whooping calls interspersed with wild roars and fearsome growls. Never before had I heard such an unbelievable cacophony of sound. Closer and closer to the lions the hyenas moved, and louder and louder grew the noise.

All at once, with fluid suddenness, the two lionesses stood up and the vibrant rattling alarm growl of the hyenas whispered all around. One of the lionesses rubbed her hind feet deliberately on the grass prior to her charge, and the hyenas turned and fled, the thudding of their heavy feet loud in the sudden silence. Both lionesses charged for at least a hundred yards, and one of them got very close indeed to old Baggage.

Then the lionesses returned to their kill and continued to feed, and for ten minutes or so the hyenas were quiet, lying half-hidden in the grass. But then the growling and the 'ooooo-whupping' began again. More and more hyenas got up and moved about restlessly. The calling got louder. Lady Astor was roaring outside my window in tones that would have made my blood curdle had I not been inside the car. Once again the Scratching Rocksters moved closer and closer to the lions, and when the foremost of them were no farther than twenty yards from the pride, the din was unbelievable. (And, through it all, Grublin slept peacefully curled up on the seat.)

But all at once the noise stopped and again the soft alarm growl vibrated in the air. The lionesses had not moved and we were puzzled until, following Bloody Mary's gaze, we saw the black-maned lion strolling towards us, followed by a group of jackals. When he was at least one hundred yards away from the nearest hyena he charged, and he kept going for a good two hundred yards – a long chase for a lion. Then he bounded over to the kill and the lionesses and cubs hurried out of his way. As he settled down to feed they lay down nearby but made no attempt to join him.

Now the jackals came into their own, for male lions, possibly through sheer laziness, are usually very tolerant towards these quicksilver scavengers. Soon there were seventeen jackals darting in and out to the kill, seizing scraps whenever the chance offered. By contrast, the arrival of the lion put an end to the aggressive demonstrations of the hyenas. Many of them, the higher-ranking individuals for the most part, left altogether, and the others settled down at a safe distance, forming a rough circle about eighty yards around the lions. There they would remain until the pride finally moved away, leaving, for the hyenas, a few bones.

It is not at all surprising that hyenas show such respect for adult male lions. One night Scratching Rocks Clan hyenas killed a wildebeest right on top of Scratching Rocks Hill, up against the tumble of grey rocks. It was at the time when the neighbouring Munge River Clan had extended its northern boundary on to the hill itself and so, as the calls of the feeding Scratching Rocksters grew louder, more and more hyenas from the Munge River Clan arrived on the scene. These moved backwards and forwards along the invisible boundary, digging at the ground in frustration. Presently, we knew, their group would be large enough, and aggressive enough, to risk a charge at the feeding Scratching Rocksters.

Both clans were making a lot of noise, and when, with startling suddenness, two lions bounded over the rocks from the darkness beyond, neither clan noticed until the enemies were right on top of the Scratching Rocksters. Then the hyenas fled, but one of them ran the wrong way in his panic and, with a lion close behind him, was trapped up against the rocks. As the lion sprang the white dust rose in the moonlight and for a moment we could not see what was happening.

When the dust settled we saw the hyena propped up by its front legs, unable to move. It was Old Gold and, as we discovered afterwards, his back had been broken. Both lions had charged off after other hyenas but soon one of them, a huge black-maned male, returned. After standing for a moment, his tail lashing from side to side, he deliberately approached the wounded hyena. Then, as Old Gold cowered away, his mouth wide open in a horrifying grimace, the lion seized his throat and slowly chewed and choked him to death. Neither Hugo nor I shall ever forget the viciousness, the seeming hatred, with which that lion killed Old Gold: indeed, for some while afterwards we were both trembling from the sudden unexpected ferocity of the attack.

But even whilst this killing was taking place, and to our utter astonishment, a hyena from the Munge River Clan calmly walked up to the temporarily deserted kill and began to feed. It had only been there for a moment when the second lion, also a full-grown male, returned to the scene and, seeing the audacious creature at the carcass, instantly charged with a loud growl. The hyena fled and actually escaped, but not before it had been severely mauled. We saw it the following day, horribly torn and hardly able to move, and we knew it was most unlikely to survive its wounds.

The lion did not eat his hyena victim: indeed, though there are several records of hyenas being killed by lions, I have only heard of one case when lions killed and then *ate* a hyena. Even hyenas, it seems, are not too keen on tackling a dead hyena – at least in the crater where food is plentiful. The body of Old Gold lay on the top of Scratching Rocks Hill, where the lion had left it, throughout the following day. In the evening several of his former clan mates sniffed at him as they walked past, but moved on. In the end it was

Mrs Brown who began the gruesome cannibalistic task of clearing away his body, and presently she was joined by her friend, old Baggage. But even then, although several other hyenas wandered up, none of them joined in until a small procession of cubs arrived from the Den of the Golden Grass at the foot of the hill. Sauce, after first rolling on the body, began to feed with gusto. Pickle instantly followed suit, and then Brindle and the other cubs joined in. Almost immediately adult hyenas began to appear from all around and took part in the feast.

In the crater, many of the hyenas have little need to scavenge – prey is plentiful and the clans are able to hunt most of their food for themselves. It is probably true to say that the higher the rank of a hyena the less need there is for it to scavenge. Bloody Mary and Lady Astor, for instance, eat their fill from the kills of their own clan almost every night. Why should they trouble to run for miles just because they see a vulture land in the distance, or hear the faint giggle of a feeding hyena somewhere in the far reaches of their territory? Why should they wait, all night, around the outskirts of a feeding pride of lions in the hope of gleaning a few bones? But the lower-ranking adults usually get much less from the clan kills. These hyenas, therefore, are far more alert to any signal from a far-off feast.

Certainly in his physical make-up the hyena is well equipped for scavenging. His ears are not only sharp, but they can accurately interpret the exact direction of a given sound; his powers of endurance are remarkable, and he can keep up a fairly fast lope for about ten miles; he has the patience to wait all night for the remains of a lion kill or to follow after a large wounded animal until it is weak enough for him to attack. In addition, his powerful teeth and jaws can crack all but the toughest bone; his digestive juices can dissolve almost anything; and he is an opportunist, willing to try anything that might serve as fuel to his metabolism – anything from leather boots to the insides of a motor car.

I wonder why it is that we humans are so quickly revolted by feeding habits that are foreign to us. The thought of snails and frogs' legs, to many Englishmen, is horrifying; Western civilisation revolts from the very idea of fried termites or soup served with live fish swimming around in a bowl. Even in my own family Hugo is horrified when I eat a kipper for breakfast, and I at the Dutch raw herrings which he loves. Small wonder then that most people are disgusted at feeding behaviour so remote from our own as that of the hyena. At first I was disgusted myself, but after a while I found that I had lost most of my squeamishness. I suppose I have become attuned to the nature of the hyena; when I watch him I switch to a different wave length. Mrs Brown so obviously relishes a steaming mouthful of entrail filled with half-digested grasses, and I observe the meal through her hyena-biased vision. Quite mouth-watering. Only if, for an instant, I imagine *myself* taking a mouthful, am I revolted. It is the same when I watch Vodka licking dry blood or drooling saliva from his mother's jowls, or Wellington urinating in the water as he drinks.

Most people, however, will continue to be revolted. And they will probably be even more revolted by the hyena's toilet. For if ever a hyena demonstrates pure bliss it is when she rolls, almost voluptuously, in some object which, to humans, is completely repulsive: like a piece of rotten gut, a dead animal, a pile of dung or, ecstasy of ecstasies, her own or her companion's vomit. Nor will it help to point out that even the most delicately nurtured drawing-room lap dog would, if it got the chance, perfume its pampered body with precisely similar odours.

Hyenas are frequently sick, and always they roll in it. It was not for some while that I realised that, for the most part, the hyena is not being sick in the normal sense of the word, but is actually regurgitating a mass of undigestible hair. Often, before or after rolling on this hair mass, a hyena picks out fragments of partially dissolved bone – when the hyena chews on them it seems that they are soft for there is no sound.

A hyena amongst a group of this kind faces quite a problem when he wants to regurgitate a hair mass. For, whilst he himself is keen to pick out the bits of bone and to roll on the remains, all his companions usually want to do the same. Once I watched Nelson, who had been resting near the Den of the Golden Grass, spit out a hair mass. Almost before it reached the ground this hair mass was covered by three rolling cubs. Nelson started to bring up a second mass, and two more cubs promptly ran up to him and hovered, waiting to roll the moment it left his mouth. Nelson paused, between heaves, to stare at them with his one good eye, heaved again, caught the mass in his teeth, rushed some distance away, dropped it, picked out a couple of bits of bone, and was just able to get his neck on to it for the start of a luxurious roll before the cubs caught up with him. Then, as Nelson rolled on the prize, one of the cubs, sniffing around, must have detected a trace of the heavy odour on its owner's muzzle and promptly rolled on that!

I remember, too, an occasion when Bloody Mary brought up a hair mass. As she started to roll on it Vodka rushed over and rolled beside her. Just then Lady Astor noticed, loped over, and also rolled – as close to the hair mass as she could get. This happened to be more or less on top of Vodka, who was then about one year old. I caught a glimpse of his puckered face, and two front paws vainly scrabbling, as he tried to extricate himself from beneath the vast weight of his mother's best friend.

Exactly why hyenas, along with so many other carnivores, like to roll on strong-smelling substances we do not know. But after all, humans (especially women) are fond of anointing their bodies with strong-smelling substances too. And considering that the foundation for so many expensive perfumes comes from the anal glands of civets, perhaps we should not be too surprised at or critical of the hyena's taste in suitable odours.

Many animals which rest above ground during the African day are pestered by flies – biting flies, stinging flies, or flies which merely irritate by tickling. The sight of a sleeping lion with hosts of insects crawling under his thighs, over his belly, around his mouth and eyes, has always horrified me. The hyena, however, escapes much of this discomfort by covering his flanks and thighs with soil. As he lies on his side he digs, with his uppermost front paw, using a scooping movement that sends the dirt upwards until, at times, his hindbody is almost hidden under a mound of earth. Then, crossing his front paws over his eyes and nose, he rests in peace.

The hyena has other ways of ensuring his comfort. If it is hot he will either rest deep down in the coolness of a den, or he will lie in mud or water. If the weather is dry he often urinates before lying down, paddling round and round with his four paws close together as he does so, and then settling luxuriously in his self-made mud.

During the rainy season the hyenas spend most of the day lying in pools of rainwater. One hot day, at noon, I drove round the den area of the Scratching Rocks Clan. First I came across the dominant male, Wellington, lying on his tummy in a puddle, his chin on his paws, his eyes closed. Farther on Mrs Brown was lying in a patch of deep oozing mud. She looked up at me, brown liquid dripping from her chin, her flared disfigured nose caked in mud. Then she turned over, with a sucking gurgling sound, and settled down again. In a neighbouring mud patch Baggage was digging vigorously, sending a shower of muddy water flying from her wallow. Then she lay down on her back with all four paws in the air. Slowly the lids drooped over her large liquid brown eyes. I drove on.

Bloody Mary and Lady Astor lay side by side at the edge of a large puddle, their legs and paws comfortably in the water. Miss Hyena wandered over to join them. She stepped daintily into the slimy pool, added a little to its volume, and then settled down, immersing her sleek shining coat in the ooze.

The hyenas of the Lakeside Clan, dry season and wet season alike, spend hours lying in the shallows of the Crater Lake. This is a soda lake, and we know, from experience, that clothes washed regularly in soda water become bleached. This probably explains why the older Lakeside Clan hyenas have such pale, almost spotless coats. The Scratching Rocksters, of course, can lie in the Munge River and the Munge Swamp during the dry season, and there they run no risk of bleaching.

The quickest dip I ever saw a hyena take was when the dominant male Wellington was 'courting' Lady Astor. At least, I think he was courting her, but we shall come to that in a moment. Certainly he was following her around, a devoted shadow. Lady Astor wandered over towards the Munge River, pausing to mark a grass tuft.

Then, whilst Wellington stopped to roll vigorously on this compelling odour, Lady Astor slipped down the bank of the stream. I heard the splash as she plunged in. Wellington hurried after her, but by the time he reached the vegetation at the edge of the water Lady Astor had reappeared, wet all over. Walking briskly, she set out across the plains. Wellington, it seemed, was in a quandary. It was very hot and dry, and I felt sure that he desperately wanted to cool off in the water. On the other hand, Lady Astor was moving rapidly away, and he certainly did not want to lose her. All at once he made up his mind. He raced down to the water and, before I could count five, was up the bank again. He had taken time to souse only one side of his rump and his belly. Nor had he taken a single mouthful of water for his lips were dry. He ran off over the plains after Lady Astor.

The sex life of the hyena remains something of a mystery to me. Hugo and I have twice come upon hyenas actually mating and we have many, many times watched the 'bowing display' which is almost certainly courtship. But we cannot prove that it is courtship for we have never seen it culminate in mating. Whenever I felt certain that, at any minute, prolonged bowing was about to bear fruit, the female and her suitor, without fail, disappeared into the reed beds of the Munge Swamp – where, as I have already remarked, a car cannot follow.

I have watched many male hyenas on a number of occasions bowing to seven different attractive females – and if it was not sex that provided the attraction, heaven knows what it could have been. One evening I found Lady Astor lying on the grass with Nelson standing some ten feet behind her. Suddenly he bowed, lowering his head until his chin almost touched the ground. Then he moved forward rapidly, bowed again, and made a series of digging movements with one paw after the other, a few feet from Lady Astor's rump. As she raised her head he shot away, tripped over a tussock of grass, and fell sprawling to the ground. Lady Astor lowered her head and Nelson got up and stood looking at her. A few moments later he repeated his bowing approach and his digging, and then withdrew quickly even though Lady Astor did not move.

There were two hours of daylight left, and during this time Nelson repeatedly went through the bowing-digging, approach-withdrawal display. When he got too close Lady Astor lunged at him, causing him to race off with his tail between his legs. In between his displaying he just lay watching the female. Occasionally he uttered a few quiet 'ooooo-whups', to which Lady Astor paid no heed. Just as darkness was falling, Lady Astor got up and wandered over to the reeds. There, with Nelson following several yards behind her, she vanished.

The next day it was Black Watch, a higher-ranking male than Nelson, who was bowing to Lady Astor. Nelson was hovering nearby, but twice, when he moved closer, Black Watch chased him away. When the sun got hot Lady Astor moved down a cool den, but Black Watch, panting in the heat, merely covered himself with a huge mound of earth near the den entrance. He was still there when Lady Astor emerged at four o'clock, and he began his bowing display all over again. At dusk the pair wandered over and vanished into the reeds.

This sort of thing went on for six more days. Each day Lady Astor was attended by a male of slightly higher rank than her suitor of the day before until, on the fifth day, she was accompanied by Wellington, the clan's number one male. By this time there was quite a group of males hanging around in the neighbourhood of the pair, but if one got too close, Wellington threatened and it retreated. That evening, as usual, Lady Astor, followed by her male, disappeared into the reeds.

The following day Wellington, when he followed Lady Astor, walked with his nose practically touching her rump. Once, when she was lying in a sphinx position, Wellington went up and actually dug at her back with one paw – and, though he backed away hastily, she did not bite towards him. Again he approached and this time he laid one front leg along her side, lowered his chest to her back and mouthed at her neck. I was convinced that, at long last, I was about to witness the consummation of the bowing display.

But Lady Astor got up and, as she walked into the reeds, I could swear that she cast a triumphant look at me over her shoulder!

For the next three days I couldn't even find the pair, and, after that, it seemed that Lady Astor was no longer attracting the males. The pertinent question is, of course, 'Did she become pregnant?' and to that the answer is no. Indeed, of the six females whom I observed as recipients of male bowing displays, only one had cubs after an interval which corresponded to the sixteen-week gestation period. On the other hand, we have so far only seen bowing displays directed to females that could be in heat – young mature females without cubs or females whose cubs have been weaned or are close to weaning. It was interesting that, in each case, it was the dominant male, Wellington, who took over from the other males after the first few days.

There was one young female whose would-be suitors were constantly repulsed, one after the other, by Bloody Mary and Lady Astor. For the four days that I was able to watch the proceedings it seemed that the two took it in turns to stay near the young female. One morning Bloody Mary, who was the chaperon of the day, lay close to the young female; Nelson and Black Watch were stretched out nearby. As the sun got hotter Bloody Mary became restless and finally she got up and began to move towards a den some forty yards away. Nelson and Black Watch sat up and watched her intently and, when she was about ten yards away, they both moved towards the young female. Just then Bloody Mary turned round and instantly rushed back. The males fled and Bloody Mary lay down near the young female again. But only for ten minutes; then, once more, she moved off towards the den. This time she stopped every few yards and looked back towards the small group. Not until she was more than twenty yards away did Black Watch get up and, looking intently at Bloody Mary, move slowly towards his female. When Bloody Mary next looked round Black Watch instantly froze and, after a moment, Bloody Mary continued on her way. Black Watch promptly started towards the female. Once more he froze when Bloody Mary looked back, but this time, after staring for a short while, the dominant female slowly walked back and, with a deep sigh, lay down in the sun again. However, five minutes later she got up and set off once more. Neither Black Watch nor Nelson moved until she had disappeared into the entrance of her den. Then they instantly rushed over to the young female with their tails curling up and began to dig at the ground and paw her back. It was only I who saw Bloody Mary's head appear at the den entrance. For a few moments she did not move; then she shot out and came running back. After this the males gave up, and presently Bloody Mary and the young female moved off together to rest in the coolness of adjacent dens.

Precisely what lay behind the constant interference of Bloody Mary and Lady Astor I cannot imagine. Perhaps they were resentful of the attention the young female was receiving, or perhaps they were trying to defend her from the advances of the males. Whatever the reason, they made a good job of their self-imposed task, and the young female did not appear to object.

Immediately before becoming attractive to the males a female hyena is liable to be mobbed. I was startled one evening to see Lady Astor crouched close to the ground, surrounded by six males. The males, amongst which were Nelson, Black Watch and Wellington, pawed at the ground and bowed. They darted forward together with their tails curled aggressively and, to my amazement, Lady Astor, the dominant, aggressive second lady of the clan, crouched to the ground with a seemingly frightened grin on her face and squealed like a crying cub. Again and again the males darted towards her and suddenly Black Watch bit the back of her neck. Instantly Lady Astor was up. She charged after Black Watch, who giggled and fled, and the other males scattered. But an hour later Lady Astor was mobbed again by an even larger group of males.

This happened the day before Nelson became her first bowing suitor. On three other occasions I saw females mobbed by males the day before the start of a series of male bowing displays. One of the females, a low-ranking individual, was bitten quite badly by several males at once, but she made up for it when, during the subsequent days, she bit no less than three of her male followers, causing one to limp for days afterwards.

Mobbing is a curious phenomenon in hyenas and seems to be set off by a variety of reasons. I have already mentioned how cubs will frequently mob a low-ranking male when he arrives near their den, sometimes chasing him right away. There are other occasions when mobbing results from a squabble between two hyenas – as the dominant one stands over the other aggressively it may be joined by other hyenas who gather round growling and biting at the unfortunate victim. Sometimes, too, the older females are mobbed whilst suckling their cubs. Once, for instance, when Baggage was peacefully nursing the twins, Bloody Mary, Lady Astor and another female suddenly approached with tails curled aggressively. They stood over Baggage uttering ferocious growls and biting at her shoulders and back. The old female crouched to the ground, squealing and grinning. Then, when the attack lessened, she moved off, followed by her two screeching cubs – for a young hyena, interrupted in his feed, makes a terrible racket. Presently Baggage lay down and the twins began to suckle again. But the others followed and once more mobbed the mother: again Baggage moved off, followed by her protesting twins. This happened once more before she was finally left in peace.

I have only one clue as to the possible explanation of such behaviour. One day when Mrs Stink was nursing her half-grown cub, her elder offspring (of whose sex I am not certain) crept up to her. He put each foot down with elaborate caution and his legs were bent as though he was trying to make himself as inconspicuous as possible. Mrs Stink didn't see him until he was almost on top of her. Then she squealed, drawing her lips back in a huge grimace, and crouched to the ground. Her small cub fled; her elder offspring stood over her growling, his tail stiffly erect, his mouth pressed threateningly against her back. Soon, however, he relaxed, sniffed and licked his mother, and then wandered away. Her small cub returned to his suckling.

Had the older offspring, perhaps, been resentful of his small sibling suckling? If so, a long-term study might reveal that the mobbing of nursing mothers may, for the most part, be initiated by a previous offspring: others may join in for the sake of joining in.

The older females, or their cubs, are also liable to be harassed by other hyenas during suckling. Sometimes this occurs when an adult wanders up to greet the nursing mother. Once, for example, when Mrs Brown was suckling Brindle, Lady Astor pushed her nose under the mother's thigh in greeting. With her muzzle she pushed Brindle out of the way: screeching, the cub returned to the nipple, and this seemed to irritate Lady Astor who growled loudly and again pushed Brindle away. The cub promptly flew into a tantrum and rushed, screeching, around his mother; Mrs Brown crouched to the ground grinning submissively, and Lady Astor stood over her growling. Then the two females licked each other and the incident ended. When the suckling mother is the higher-ranking individual in such an incident it is she who growls and, if the subordinate hyena persists in its attempt to greet her, it is usually chased away by mother and cub together.

On another occasion, however, when Lady Astor herself was suckling eighteen-month-old Miss Hyena, a young adult female approached, pushed Miss Hyena from the nipple and, with bristling tail, chased the big cub twice around her mother. Lady Astor simply lay and watched. Miss Hyena then resumed her suckling, but after a moment the young female again pushed the cub away and chased her, screeching loudly, around her mother. When Lady Astor got up and walked away, Miss Hyena following, the young female went with them and, as the cub again began to nurse, very slowly pushed her muzzle between Miss Hyena and her mother. Then she placed one paw between them; then the other; then she lay down, almost on top of Miss Hyena, who calmly continued to suckle.

I have watched similar behaviour on many occasions, and I have also seen young cubs, when following their mothers, headed off again and again by young adults. I have a strong suspicion that the young adults, in many cases, are in fact the older offspring of the mothers concerned. Indeed, the young female who pushed Miss Hyena from her mother does look amazingly like Lady Astor.

Hyena cubs are not normally weaned until they are at least eighteen months old, and the weaning period, which may extend over several months, is sometimes a stressful time for both mother and cub. Lady Astor's daughter, Miss Hyena, was a real problem to her mother at this time. She was about one and a half years old, and during the three weeks that we were at the crater scarcely a day passed when I did not see Miss Hyena fly into at least one weaning tantrum. Once when she was prevented from suckling, she first backed away squealing – a harsh, grating long drawn out sound which goes on and on and on. Then she rushed back to Lady Astor and, with her legs bent so that she was moving along with her belly scraping the ground, she rushed round and round her mother, squealing and grinning with her tail erect. Again she tried to nurse, again she was repulsed by a quick nip, and again she rushed off squealing. Lady Astor walked away, followed by a squealing daughter. Ten minutes later she permitted Miss Hyena to suckle. And so it went on, day after day, the tantrums getting worse and worse, and often mother and daughter biting each other.

Three days before we left the crater I saw Lady Astor, for the first time, prevent her daughter from suckling even for a minute, despite the fact that Miss Hyena persisted for nearly an hour. Time and again the large cub lay down, very, very cautiously, and pushed her nose towards her mother's nipples; time and again Lady Astor turned and bit at her. Finally Lady Astor stood up and went round and round after her daughter in a tiny circle, biting at her neck and back again and again. How long the battle of wills might have persisted, but for the timely arrival of another cub, I cannot imagine. As it was, Miss Hyena abandoned her attempts in order to enjoy a rough and tumble, and Lady Astor was left in peace.

Most carnivores suckle their young for a few weeks only: by contrast, the eighteen or more months during which a hyena cub nurses seems amazingly long. It is, however, necessary, for a mother hyena neither leads her cubs to kills, nor does she regularly bring back food to the den. The growing youngster, therefore, depends on maternal milk for much of its nourishment.

Sometimes mother hyenas do carry bones back to their dens, and whilst they themselves may then chew on the trophies for a while, they nearly always permit their own cubs to share with them. Often, indeed, it seems that they take bones back only for the enjoyment of their offspring. Once, for instance, Baggage carried a large piece of backbone to the den where, a few hours before, she had left the twins. She put the bone down, showered it with earth as she dug at the burrow entrance, and then called softly. No cubs appeared for the twins had just wandered over to a new den. For the next twenty minutes Baggage trailed from den to den with the bone, digging and calling at each, until she finally caught up with the twins. Then she dumped her offering beside them and lay to rest, moving only to back up the threats of her cubs when another youngster tried to share their bone.

Spoils such as this, however, seldom offer much in the way of nourishment for the cubs and so, until a youngster is large enough to compete with other scavengers around carrion, until its jaws are strong enough to crunch up big bones and, above all, until it is able to take its place with the rest of the clan around a kill, its mother's milk is all-important. Bound up in these facts lies the answer to the riddle of why some hyena cubs grow much faster than others: it depends, to a very large extent, on the social status of the mothers. For a high-ranking female gets more and better food from the clan kills than a low-ranking one and so, probably, can produce more and richer milk. In addition, the cub with a high-ranking mother usually joins in the kills of his clan at an earlier age than cubs with subordinate mothers.

Thus Brindle, a single cub, was much larger at twenty months of age than the twins, who were a couple of months older at the time. Their mothers, Mrs Brown and Baggage, were roughly equal in social status, but presumably Brindle got more milk than each of the twins who had to share their mother's supply. On the other hand, the twins grew faster than another single cub, of the same age, whose mother held a social position well below that of Baggage.

The most spectacular example of the rapid development of a cub with a high-ranking mother lies, as might be expected, in the growth of Vodka, surviving cub of the clan's leading female. Vodka, as I have already described, attended clan kills at an exceptionally early age, and he normally got large quantities of food from them. When he was only one year old he was of similar size to cubs six months older than himself; moreover, Bloody Mary was actually able to wean him then, half a year earlier than most other mothers. Nor did I ever see Vodka go into a weaning tantrum when his mother rejected him: presumably he was so well nourished that milk, by then, was merely a luxury.

Our hyena study is in its infancy, yet we have some indication of the relationship which may develop between mothers and their young adult offspring. Mrs Brown and her elder son, Master Beige, have, on the whole, very little to do with each other, and when they do interact their behaviour is often aggressive. The relationships between two other females and their young adult sons is similar. By contrast, Lady Astor and Miss Hyena (who is now about four years old) frequently move about in the same group, feed together at clan kills and lie close together when resting. When Lady Astor is lying down and her daughter wanders up to greet her, it often looks as though Miss Hyena is about to suckle: she flops down with her muzzle close to her mother's nipples. And, just as she did when Miss Hyena was a little cub, Lady Astor usually lays a front paw companionably over her adult daughter's flanks. It was interesting to find that Vodka, since being weaned, has often greeted and lain with his mother in exactly the same way. Perhaps Vodka is also a female.

From the scanty knowledge available at present it seems at least probable that the status of the mother will shape the social position of her offspring within the clan. Miss Hyena, for instance, is already a high-ranking individual in her own right, even in the absence of Lady Astor. If it is also true that a high-ranking mother is likely to raise a high-ranking son, then her status, indirectly, may affect his subsequent territorial behaviour, for the most strictly territorial males are usually those of high rank. I have never seen Wellington, the number one male, penetrate a neighbouring clan's territory except in the thick of his own clan – during a border skirmish, for instance. But low-ranking, one-eyed Nelson quite often moves into neighbouring hunting grounds on solitary foraging expeditions. Indeed, I have seen him actually resting beyond the boundary of Scratching Rocks Clan territory. It is, of course, an advantage to a male such as Nelson to have available as large an area as possible for his foraging expeditions: he is subordinate even to the low-ranking females and, together with some of the cubs old enough to take part, comes off worst at the kills of his clan. However, the whole question of the territorial behaviour of male hyenas is still far from clear since, as we shall see presently, even quite high-ranking young males may become members of two different clans.

It is possible that many hyenas under about twelve to eighteen months old may be, to some extent, free from the clan restrictions which affect most of their elders. Once, when parties of Scratching Rocks and Munge River hyenas were threatening each other over a wildebeest kill, a Munge River cub calmly walked up to a Scratching Rocks cub and the two greeted each other in a most friendly way. And I have followed three different Scratching Rocks cubs, separately, on fairly long foraging excursions well beyond the boundary of their own territory. As a cub, destined to become a high-ranking male, grows older it may be his active participation in the marking parties and border skirmishes of his clan which shapes his more strictly territorial behaviour.

Particularly fascinating is the position of the male hyena who can boast membership of two different clans. Such a male may hold a relatively high rank in the clan of his birth, yet he sets out, apparently purposefully and certainly persistently, to establish relationships with individuals of a neighbouring clan. Eventually, if he is successful, he will be tolerated not only as he wanders through the hunting grounds of his neighbours, but also at their kills. At the same time he will retain his right to participate in the social life and the kills of his own clan.

We have been fortunate enough to observe the gradual acceptance of a young Scratching Rocks male into the Lakeside Clan. But the extraordinary position of this male, Quiz, cannot be fully appreciated without a more detailed knowledge of the normal relationship between the two clans.

Members of both clans frequently patrol and scent-mark the boundary between the two territories which runs for just under a mile across the open plains. Some points along this invisible line are regularly marked by groups from both clans. A clan marking party may consist of a few individuals or as many as thirty hyenas of both sexes; we have not so far seen nursing cubs take part in such expeditions.

A typical Scratching Rocks Clan marking party was initiated by Bloody Mary and Lady Astor; both dominant females were accompanied by their offspring, Vodka and Miss Hyena. Wellington, the number one male, was in close attendance. As this small group set off from the den, where they had been resting during the evening, Bloody Mary and then Wellington gave series of 'ooooo-whups'. They moved faster as they approached the territory of the Lakeside Clan, and gradually more and more Scratching Rocksters joined the party. Thirty yards or so from the boundary the leaders began to run, their tails curled over on to their rumps. The others followed suit. Suddenly Bloody Mary and Lady Astor stopped, jostling each other to get a good smell from the patch of tall grass where, just the night before, I had watched a Lakeside marking party leaving the scent of their clan. Soon the other Scratching Rocksters caught up, and the higher-ranking individuals crowded into a tight-packed group, each one trying to push aside the nose of his or her neighbour in order to get a better whiff. The lower-ranking hyenas rushed about on the outskirts, and the whole group showed intense excitement.

Bloody Mary was the first to deposit Scratching Rocks Clan scent. With her hind legs bent she slowly moved forward so that a long stem of grass was drawn between her hind legs. I could see her scent glands protrude in a bulge just above her anus as she left a smear of the strong-smelling secretion on the stem. Lady Astor was close behind and she marked the grass the moment Bloody Mary had finished. Then, whilst the two dominant females marked another stem, the rest of the group queued up to follow their example. Wellington, after marking one of the stems, urinated, digging with his front paws where the stream hit the ground and splashing himself and his surroundings. Presently other males did the same.

Before the lowest-ranking hyenas of the group had managed to reach and mark the chosen grasses, Bloody Mary and Lady Astor had set off briskly for the next marking place, their tails still curled, their sides touching. The other hyenas, having finished their marking, hurried after their leaders. And so it went on along the border. At no time were any hyenas of the Lakeside Clan sighted and, after also marking part of the boundary between their territory and that of the Munge River Clan, the Scratching Rocksters dispersed, wandering back into their own hunting grounds.

On those occasions when marking parties from Scratching Rocks and Lakeside Clans met each other there were always skirmishes, when the Scratching Rocksters charged at the Lakesiders, and vice versa, each clan becoming the aggressor when the other penetrated too far into hostile territory. These skirmishes sometimes resulted in bleeding ears or slight limps; only once, as described at the beginning of this chapter, did we see a marking party actually catch and maul a neighbouring clan member.

Most of the aggressive incidents between clans that we have watched have been over food. Once, for example, hyenas of the Scratching Rocks Clan killed a wildebeest within thirty yards of the boundary between their territory and that of the Lakesiders. As the sounds made by the Scratching Rocksters as they squabbled over the meat grew louder, so more and more Lakesiders came running to the scene. After some five minutes the Lakesiders, frustrated and hungry, had worked themselves into a frenzy of aggression. When they charged, in a tight group, at the Scratching Rocksters, the latter fled: their leaders were full and contented and, obviously, not aggressive enough to withstand the onslaught. After another five minutes, however, the higher-ranking Lakesiders were equally replete and, when the Scratching Rocksters, their frustration having built up as they watched other hyenas feeding, charged in turn, the rival clan fled. That kill changed clans five times before it was consumed.

We have seen incidents of this sort many times: on two occasions hyenas from both clans actually fed from either end of the carcass at the same time, but only for a few moments, and to the accompaniment of the most ferocious growls and roars.

The situation is very different if a kill is made well within the territory of a neighbouring clan. Then the hunters, however hungry, and even if they have scarcely had a bite from their kill, usually retreat precipitously when faced by a group of territory owners. Once, when the Lakesiders made a kill in Scratching Rocks territory, they had fed for some five minutes before a group of Scratching Rocksters raced to the scene. The intruders fled, but one of them, a young male, failed to escape. Quite a large group of high-ranking Scratching Rocksters ignored the dead wildebeest and concentrated on savaging their victim. They mauled him in the most horrible manner and only left him when he was completely crippled and unable to move. By morning he had died of his wounds. On subsequent occasions we found other hyenas mauled in the same way: each time they were fairly young males. It is not impossible that, like young Quiz, they were individuals who had just gained acceptance into a neighbouring clan but that their new friends, in the heat of battle, failed to recognise the fact.

This territorial ferocity makes the story of Quiz particularly fascinating. Quiz is not only high-ranking amongst his age group – he is about four years old – but he dominates a number of the younger females of the Scratching Rocks Clan as well. Why should he expose himself to the danger of trying to enter the society of the aggressive Lakesiders?

It was within a somewhat dramatic framework that first I saw him make overtures to individuals of the Lakeside Clan. Scratching Rocksters had killed close to the territorial boundary between the two clans, just within their own home ground. Within a few minutes there were thirty-nine Scratching Rocks hyenas around the kill, and soon many of them had blood-soaked faces and necks, scarlet in the lurid light of the rising red sun. The Lakesiders gathered fast, uttering growls and low whoops, digging in the white dust just within their own boundary.

Suddenly I heard the whispering rattle of the alarm growl and all the hyenas fled as two lions bounded towards the kill. Then, as the lions fed, the individuals of the two clans ranged themselves on either side of the invisible scent boundary; hyenas of both groups paced back and forth with curled tails, or lay staring towards the lions. All at once a lone figure walked out into the corridor between the clans. It was Quiz. When he got to within twenty yards of the nearest Lakeside hyenas he stopped and stood with his head up and his tail down, staring towards them. After a few moments a young hyena, about the same age as Quiz, walked towards him. As it approached, Quiz darted away for a few steps, his tail tucked between his legs. But then he stopped and held his ground as the Lakesider greeted him, pushing itself under his chin, rubbing against his chest. Then the two sniffed under each other's raised hind legs. I felt sure they had met before.

When Quiz looked up from this greeting he saw a line of eight Lakesiders walking towards him, all with aggressively curled tails. After staring at them for a moment he turned and ran back to his own clan. He went straight up to Bloody Mary and greeted her, sniffing under her leg, rubbing against her, nosing and licking her mouth and, all the time, he wagged his tail vigorously with crouched hindquarters, and made jerky bobbing movements of his head – all signs of a submissive hyena. Then he moved around greeting other hyenas, one after the other. It was as though he sought to propitiate the members of his own clan after his alien contact.

Quiz had been on Lakeside soil for a total of thirteen minutes, and this, I thought, was the end of the incident. But it was not. Twenty minutes after returning to his own clan he once more moved away across the territorial boundary. By now there were only five Lakesiders left, one of which was a very high-ranking male. None of the five moved as Quiz approached, and after standing for a moment he went right up to one of them and initiated a greeting. When the dominant male got up and approached, his tail curled, Quiz darted away, but only for a few yards. Then he stopped, turned round, and walked towards one of the others.

For half an hour Quiz made repeated overtures to those five Lakesiders, and several times established contacts with three of them. During each of these greetings the Lakesider held its tail curled stiffly upwards or hanging and relaxed, whilst Quiz kept his tucked submissively between his legs and gave frequent nervous bobbing movements of his head. And each time the dominant male approached to within ten yards or so, Quiz hastily retreated. The Lakesiders, following the example of the dominant male, frequently marked a certain clump of grass near Quiz, but Quiz, although he approached and sniffed the spot, did not mark it himself.

Suddenly the two lions left the kill. The five Lakesiders raced towards the remains, but so did some twenty Scratching Rocksters, and the five turned and fled. To my surprise Quiz, who had stood watching as the Lakesiders were chased off, made no attempt to join his own clan at the kill, but moved back into Lakeside territory and lay down, quite close to the routed five.

When most of the carcass had been divided and only a group of six Scratching Rocksters growled around the remains, the Lakesiders again advanced but were quickly chased off by the six Scratching Rocks hyenas. This time Quiz joined his own clan in the charge against the Lakesiders; but when the others returned to the carcass he did not follow. Instead, as though made bold by renewed contact with his own clan, he ran to the Lakesider marking place and, for the first time, with his tail curled up, he added the scent of the Scratching Rocks Clan. This, it seemed, could not be tolerated. As one, the five Lakesiders instantly rushed at Quiz and he fled back over the boundary to his own clan. And there he remained, having spent a total of one hour and fifteen minutes on Lakeside territory and made contact with four different individuals of the neighbouring clan.

Subsequently, Hugo twice saw Quiz actually feeding with the Lakeside Clan at their kills, and whilst he did not get large amounts of the food, he did no worse than many of the Lakesider males themselves. After one of these kills, Quiz walked with a group of

Lakesider males towards the communal dens of that clan, but before he got there he stopped, watched his new friends move on, and returned to his own territory. Another time Quiz suddenly left a Scratching Rocks marking party and ran off by himself into the hunting grounds of the Lakeside Clan. He seemed worried, and kept stopping and listening, but he went on nevertheless. As darkness fell we left him, a rather lonely-looking figure, still penetrating deeply into Lakeside territory.

That is one hyena picture I shall never forget, one more to add to my store. For although it is several months since we last worked in the Ngorongoro Crater, I need only close my eyes for the hyenas to be all around me again. I can see Mrs Brown, with her chewed-off disfigured nose, lying by her den and watching Brindle as he gnaws on part of a wildebeest skull. As an older cub approaches, Brindle struggles to pick up the skull. At last he gets a grip and as he walks, staggering under the weight, the two horns sweep out on either side of his face like a monstrous moustache. And there are the twins shambling along, leading a small procession of cubs on an evening stroll. Sauce, closely followed by Pickle, approaches a bull wildebeest. The two run forward bravely with their small tails curled, whilst the other cubs follow more slowly, their tails only horizontal. The wildebeest stands and stares at them and, one by one, the cubs stop and stare back. Suddenly the wildebeest snorts and tosses his head, and the cubs, their bluff called, turn and race away with their tails tightly tucked between their legs. As they run towards the safety of their den, Baggage arrives, her pendulous belly brushing the tips of the golden grass stems. She pokes her head down the burrow and digs, seemingly oblivious to the fact that her cubs are running up behind her. As they arrive she envelops them both in a cloud of dust.

And there is Countess Dracula, with her split sneering lip, moving one of her tiny cubs. Two bat-eared foxes are trotting along beside her. Their huge ears prick forward as they stare curiously at the black body swinging from the hyena's mouth. When Countess Dracula vanishes into her new den the foxes peep down and then, hunting insects, slowly wander away.

There is one beautiful, vivid picture of Nelson too. He is outlined by the setting sun with a fringe of pure gold as he walks, knee-deep, through the shining pinkness of flowering grasses. 'Ooooo-whup! Ooooo-whup!' he calls softly to himself, whilst above him the dancing lake flies are a thousand silver specks against the darkness of the crater rim.

I can see Bloody Mary and Lady Astor, side by side, at a kill, their faces and necks bright red with blood in the spotlight. Miss Hyena is pressed close to her mother, and Vodka is right under Bloody Mary's belly, lying on part of the carcass, gorging himself. Beyond the circle of light the eyes of the low-ranking hyenas prowling on the outskirts shine like stars. And the growls and the whoops and the roars and the chuckles evaporate, one after the other, into the darkness.

One last picture. Vodka, with a large juicy bone in his mouth, is following his mother from a morning kill. He stays close to her for Miss Hyena is walking beside him and casting longing glances at his bone. When Bloody Mary lies down near her friend, Lady Astor, Vodka lies too. But Vodka is very thirsty. With his bone in his mouth he walks to a nearby pool of water. Laying down his bone, he lowers his head to drink but, from the corner of his eye he sees Miss Hyena cautiously approaching his bone. He snatches it up and returns to the security of his mother. A few moments later, however, his thirst becomes unbearable. Again he moves to the water and sets down his bone: again as he is about to curl his tongue for the first cool sip he has to turn to snatch the bone away from Miss Hyena. He stands, the bone in his mouth, looking from the water to Miss Hyena, from Miss Hyena to his mother. Bloody Mary is asleep. Twice more Vodka's attempts to drink are frustrated; the next time Miss Hyena gets the bone. As she runs to eat it by Lady Astor, Vodka stares after her, but makes no attempt to follow. He drinks his fill and then, his sides bulging from all that he has eaten, he returns to lie pressed close to Bloody Mary.

The last time we were in the crater, Bloody Mary was pregnant again. By now, if all has gone well, her cubs will be born. I wonder whether she has again chosen, as her first nursery, the large burrow in the disused termite mound? And how many small black faces with misty grey-blue eyes are peering from the den entrance at the greenness of the rainy world? Most of all, I wonder how Vodka, who has, for so long, been the constant companion and only child of his mother, has reacted to the arrival of his new siblings. Soon, I hope, Hugo and I will go back and see.

Epilogue: Looking Ahead

For over two years we have studied hyenas, jackals and wild dogs. Hour after hour we have watched them and followed them on their day-to-day activities. We have spent days and nights with them, watched their youngsters grow up, shared in many of their tragedies and pleasures. In short, we have tried to understand why they live as they live.

We have now begun to study the lives of lions, leopards and cheetas for a new book, *The Stealthy Killers.* Two years of work lie ahead. But however fascinated we become with the different individuals amongst the great cats, we shall not forget the others. We plan, for instance, to make frequent visits to Ngorongoro Crater to find out what is going on in the clan of the Scratching Rocks hyenas. We want to follow the development of Bloody Mary's cub, Vodka – we want to discover, for one thing, his sex. And to find out who will take over leadership of the clan when Bloody Mary becomes too old, or dies. And we want to see how Lady Astor, second female of the clan, reacts when her own pretty daughter, Miss Hyena, produces her first cubs. Particularly we are interested in following the progress of young Quiz, the two-clan male. For all we know he may, in his old age (when he is thirty years or so), be an accepted member of several of the crater's hyena clans.

We are equally anxious to find out what is happening to our golden jackals. In fact, we are planning to go to Ngorongoro in a few weeks' time for, by then, Cinda, daughter of Jason and Jewel, should have her own cubs. Will she, we wonder, rear her offspring in the dens where she herself spent her childhood? The memories of that period are, for us, many and vivid: the boisterous tumbling play of Rufus and Nugget and their sister Amba; the way in which Cinda, the runt, so frequently curled up by herself; Cinda in the talons of the eagle, and her shrill screaming as she hurtled to the ground; Jewel, their mother, pouncing on cub after cub, bowling it over, and then grooming it until finally it escaped to join the games of its siblings; Jason darting in and out of lion and hyena kills for titbits; Jason battling with a snake; Jason, alone, challenging and driving off a huge lappet-faced vulture from the food of one of his cubs.

Surely Cinda is not the only surviving member of that vigorous family? We shall stay long enough at the crater to make another search, for this is the time, when the jackals so often have their cubs, that we have the best chance of finding old Jason and Jewel, or Rufus, Nugget and Amba – if any of them are still alive.

At present we are camped at Ndutu tented lodge with our friend George Dove. George became too good a friend for us to continue camping on the opposite side of the lake: we made so many trips to his camp for supplies and water – and quick chats; he drove over to us so often to visit his special friend, Grublin. And so we compromised, and, with our assistants, sleep and work in our own tents, but eat with George and share his facilities. Now we can chat whenever there is time – and without wasting a drop of petrol.

As I write, the migration of wildebeests, zebras and gazelles is all around the lake, together with its retinue of lions, cheetas, hyenas and jackals. And, of course, the wild dogs. Every time a pack of dogs is sighted (George's tourist visitors are incredibly helpful to us) we drive quickly to the place to make observations and take photographs. We already have recognition pictures of one hundred and sixty three different individuals pasted into our file. Last year, when we came to Lake Legaja to study wild dogs, the packs were few and far between; the days when exciting information was gleaned were scarce. This year one or more packs are seen once every three or four days.

And last week we found the Genghis pack – we still call it the Genghis pack even though the old leader has gone. The pups are over half grown now and, although one has disappeared, the others are in excellent condition. We met up with the pack at a most opportune time, for the dominant female, Havoc, was in heat. The relationships which she established with old Yellow Peril, and with the dominant Swift and Baskerville, during the three days that we were able to keep up with the pack, were fascinating beyond words. We even saw a pack fight during which Yellow Peril was mobbed by all the others, adults and pups alike. And during that time we heard the dogs utter strange and extraordinary sounds which we had never heard before. In fact, the information we obtained during those three days is not only new to science, but will provide the nucleus for a film, and possibly an entire book devoted to these fascinating animals.

We lost the pack during the darkness of night, and although we searched with three cars for the whole of the next day we could not find them. Nor did we see them the following day when a pilot friend, who had brought some tourists to George's camp, offered to fly us over the area. Grublin came with us, perched on George's knee, thrilled with the aeroplane and the sight of the migrating herds below. But although we zigzagged back and forth over the plains for two hours and saw lions and cheetas, hyenas and jackals, we saw no glimpse of any wild dogs.

Where have they gone, and why? They have left an area thick with animals, offering an abundance of food. Perhaps they would cry, with the runners in C. H. Sorley's poem: '. . . we run because we must . . .' or '. . . we run because we like it, Through the broad bright land'.

However, whilst, with our assistants, we observe the lions and the cheetas as they roam through the herds of the migration, we shall continue to search for the pack. And, if we have not located them in eight weeks' time, we shall hire another aeroplane and fly, backwards and forwards over the Serengeti, until we do find them. For by then Havoc should produce her pups. Will Black Angel be as fascinated by Havoc's pups as she was by Juno's? I can still see her, surrounded by the eight round-bellied youngsters, licking them, carrying them around, utterly captivated by the tiny creatures. If so, it may be that her determination to keep the other dogs away from the new pups may boost her social rank and reinstate her as female number two of the pack – for when we met them she was still bottom female.

As yet the characters of the big cats we are watching are undefined, except for those of a mother leopard and her almost adult cub whom, with our assistants, we have been able to watch, almost continuously, for four months. But the lazy male lions and the hard-working often harassed lionesses of the pride are, so far, just lions. And the mother cheeta and the seldom seen males are, as yet, even more wraith-like in character. But soon, as we get to know their habits, begin to understand the motives which direct their existence, start to appreciate the traits of the different individuals, their personalities will become as real and vivid as those of Bloody Mary, Cinda or Black Angel.

By that time, since we are still intensely concerned with the doings of some fifty wild chimpanzees on the shores of Lake Tanganyika, it seems that our life work will be cut out, for we shall be trying to keep up with the goings on of over a hundred individual animals of seven different species spread out across the vast spaces of Africa's Tanzania.

Bibliography

As this book has been written for the general public those scientists and other authors whose work we have referred to have not always been specifically mentioned in the text. The following publications have all been of value to us during our research. Titles marked with an asterisk * are those recommended for the non-scientist as providing additional background information on the species described in this book, together with some of their closest relations.

*Allen, D. L. and L. D. Mech (1963). Wolves versus Moose on Isle Royale. *Nat. Geog. Mag.* 123, 200–219.

Brown, L. E. (1966). Home Range and Movements of Small Mammals. *Symp. Zool. Soc. Lond.*, No. 18, 111–142.

Burrows, R. (1968). Wild Fox. David & Charles Ltd., Newton Abbot, Devon.

Cowie, M. (1966). The World of Animals: The African Lion. Arthur Barker Ltd., London & Golden Press, New York.

*Estes, R. D. (1967). Predators and Scavengers: Stealth, pursuit and opportunism among carnivores on Ngorongoro Crater in Africa. Parts I and II. *Natur. Hist.* 76(2): 20–29; and 76(3): 38–47.

*Estes, R. D. and J. Goddard (1967). Prey selection and hunting behaviour of the African wild dog. *J. Wildl. Manage.* 31(1): 52–70.

Fuente, F. R. de (1970). La aventura de los lobos. *La Actualidad Española*, 941: 29–64.

Grafton, R. N. (1965). Food of the Black-backed Jackal. In: Proceedings of a symposium on African mammals. *Zool. Africana* 1(1): 41–53.

Jewell, P. A. (1966). The Concept of Home Range in Mammals. *Symp. Zool. Soc. Lond.*, No. 18, 85–109.

Jacobi, E. F. and A. C. V. van Bemmel (1968). Breeding of Cape Hunting Dogs (Lycaon pictus (Temminck)) at Amsterdam and Rotterdam Zoo. *Der Zoologische Garten*, Band 36, Heft 1–3.

Jordan, P. A. and P. C. Shelton and D. L. Allen (1967) Numbers, Turnover and Social Structure of the Isle Royale Wolf Population. In: Animal Behaviour Society: Ecology and behaviour of the wolf. *Amer. Zool.* 7(2): 233–252.

Joslin, P. W. B. (1967). Movements and Home Sites of Timber Wolves in Algonquin Park. In: Animal Behaviour Society: Ecology and behaviour of the wolf. *Amer. Zool.* 7(2): 279–288.

Kleiman, D. (1966). Scent-Marking in the Canidae. *Symp. Zool. Soc. Lond.*, No. 18: 143–165.

Kleiman, D. (1967). Some Aspects of Social Behaviour in the Canidae. In: Animal Behaviour Society: Ecology and behaviour of the wolf. *Amer. Zool.* 7(2): 365–372.

Klingel, H. (1967) Soziale Organisation und Verhalten freilebender Steppenzebras. Z. Tierpsychol. 24, 580-624. Zeitschrift für Tierspsychologie.

Kruuk, H. (1966). Clan-system and feeding habits of spotted hyenas (*Crocuta crocuta* Erxleben). *Nature*, 209(5029): 1257–1258.

*Kruuk, H. (1968). Hyenas: The Hunters Nobody Knows. *Nat. Geog. Mag.* 134(1): 44–57.

Kühme, W. (1964). Die Ernahrungsgemeinschaft der Hyanenhunde (*Lycaon pictus lupinus*, Thomas, 1902). *Naturwissenschaften* 51(20): 495.

Kühme, W. (1964). Uber die soziale Bindung innerhalb eines Hyanenhund-Rudels. *Naturwissenscnaften* 51: 567–568.

Kühme, W. (1965). Communal food distribution and division of labour in African hunting dogs (*Lycaon pictus lupinus*). *Nature* 205: 443–444.

Lawick-Goodall, J. van and H. van Lawick (1966). Use of Tools by the Egyptian Vulture (*Neophron percnopterus*). *Nature* 212(5069): 1468–1469.

Lawick-Goodall, J. van and H. van Lawick (1967). Tool-using bird: the Egyptian Vulture. *Nat. Geog. Mag.* 133(5): 631–641.

*Leger-Gordon, D. St. (1951). The Way of a Fox. John Murray, Lond.

Lockie, J. D. (1966). Territory in Small Carnivores. *Symp. Zool. Soc. Lond.*, No. 18, 143–165.

*Lorenz, K. (1955). Man Meets Dog. Houghton-Mifflin, Boston.

Maxwell, G. (1960). Ring of Bright Water. Longmans, Lond.

*Murie, A. (1962). On the Track of Wolves. *Nat. Hist. N.Y.* 71(8): 29–37.

Österholm, H. (1964). The Significance of Distance Receptors in the Feeding Behaviour of the Fox. *Vulpes vulpes L.* Acta Zoologica Fennica 106: 1–31. (*Acta. Zool. Fen.*)

Ozoga, J. J. and E. M. Harger (1966). Winter Activities and Feeding Habits of Northern Michigan Coyotes. *J. Wildl. Manage.* 30(4): 809–818.

Pimlott, D. H. (1967). Wolf Predation and Ungulate Populations. In: Animal Behaviour Society: Ecology and behaviour of the wolf. *Amer. Zool.* 7(2): 267–278.

*Pimlott, D. H. World of the Wolf. 1969. New York.

Pournelle, G. H. (1965). Observations on the birth and early development of the spotted hyena. *Jour. Mamm.* 46: 3.

Rabb, G. B. and J. H. Woolpy and B. E. Ginsburg (1967). Social relationships in

a group of captive wolves. In: Animal Behaviour Society: Ecology and behaviour of the wolf. *Amer. Zool.* 7(2): 305–311.
Schenkel, R. (1967). Submission: its features and function in the wolf and dog. In: Animal Behaviour Society: Ecology and behaviour of the wolf. *Amer. Zool.* 7(2): 319–329.
Scott, J. P. (1967). The evolution of social behaviour in dogs and wolves. *Ibid.* 373–381.
Seitz, v. A. (1954). Beobachtungen an handaufgezogenen Goldschakelen (*Canis aureus algirensis*, Wagner, 1843). *Z. Tierpsychol.* 16: 747–771.
Snow, C. J. (1967). Some observations on the behavioural and morphological development of coyote pups. In: Animal Behaviour Society: Ecology and behaviour of the wolf. *Amer. Zool.* 7(2): 353–355.
Tinbergen, N. (1964). The fox and its food store. *Lerende Nat.* 67: 73–79.
Woolpy, J. H. (1968). The Social Organisation of Wolves. *Nat. Hist.* 77(5).
Wright, B. S. (1960). Predation on big game in East Africa. *J. Wildl. Manage.* 24: 1–15.
*Young, S. P. and E. A. Goldman (1966). The Wolves of North America. Parts I and II. Dover Publications Inc., New York.

Index

main track or dirt road - others change yearly
disputed territory with
Table Mountain Clan
grass and low shrub
drinking
place
Table Mountain
Scratching
reed
beds
lake
Jason's range
marsh
Jackals
Hyenas
Crater Lake